Lecture Notes in Mathematics 1825

Springer
Berlin
Heidelberg
New York
Hong Kong
London
Milan
Paris
Tokyo

J. H. Bramble A. Cohen W. Dahmen

Multiscale Problems and Methods in Numerical Simulations

Lectures given at the
C.I.M.E. Summer School
held in Martina Franca, Italy,
September 9-15, 2001

Editor: C. Canuto

Fondazione
C.I.M.E.

Springer

Editor and Authors

Claudio Canuto
Dipartimento di Matematica
Politecnico di Torino
Corso Duca degli Abruzzi 24
10129 Torino, Italy

e-mail: claudio.canuto@polito.it

James H. Bramble
Mathematics Department
Texas A&M University
College Station
Texas TX 77843-3368
USA

e-mail: james.bramble@math.tamu.edu

Albert Cohen
Laboratoire Jacques-Louis Lions
Université Pierre et Marie Curie
175 rue du Chevaleret
75013 Paris, France

e-mail: cohen@ann.jussieu.fr

Wolfgang Dahmen
Institut für Geometrie
und Praktische Mathematik
RWTH Aachen
Templergraben 55
52056 Aachen, Germany

e-mail: dahmen@igpm.rwth-aachen.de

Cataloging-in-Publication Data applied for

Bibliographic information published by Die Deutsche Bibliothek

Die Deutsche Bibliothek lists this publication in the Deutsche Nationalbibliografie; detailed bibliographic data is available in the Internet at http://dnb.ddb.de

Mathematics Subject Classification (2000): 82D37, 80A17, 65Z05

ISSN 0075-8434
ISBN 3-540-20099-1 Springer-Verlag Berlin Heidelberg New York

Springer-Verlag Berlin Heidelberg New York a member of BertelsmannSpringer
Science + Business Media GmbH

http://www.springer.de

Typesetting: Camera-ready TEX output by the authors

SPIN: 10953471 41/3142/du - 543210 - Printed on acid-free paper

These Lecture Notes are dedicated to the victims of the brutal attacks
of September 11, 2001, including all who were affected. All of us who
attended the C.I.M.E. course, Americans and non-Americans alike,
were shocked and horrified by what took place.
We all hope for a saner world.

Preface

The C.I.M.E. course on "Multiscale Problems and Methods in Numerical Simulation" was held in Martina Franca (Italy) from September 9 to 15, 2001. The purpose of the course was to disseminate a number of new ideas that had emerged in the previous few years in the field of numerical simulation, bearing the common denominator of the "multiscale" or "multilevel" paradigm. This takes various forms, such as: the presence of multiple relevant "scales" in a physical phenomenon, with their natural mathematical and numerical counterparts; the detection and representation of "structures", localized in space or in frequency, in the unknown variables described by a model; the decomposition of the mathematical or numerical solution of a differential or integral problem into "details", which can be organized and accessed in decreasing order of importance; the iterative solution of large systems of linear algebraic equations by "multilevel" decompositions of finite-dimensional spaces.

Four world leading experts illustrated the multiscale approach to numerical simulation from different perspectives. Jim Bramble, from Texas A&M University, described modern multigrid methods for finite element discretizations, and the efficient multilevel realization of norms in Sobolev scales. Albert Cohen, from Université Pierre et Marie Curie in Paris, smoothly guided the audience towards the realm of "Nonlinear Approximation", which provides a mathematical ground for state-of-the-art signal and image processing, statistical estimation and adaptive numerical discretizations. Wolfgang Dahmen, from RWTH in Aachen, described the use of wavelet bases in the design of computationally optimal algorithms for the numerical treatment of operator equations. Tom Hughes, from Stanford University, presented a general approach to derive variational methods capable of representing multiscale phenomena, and detailed the application of the variational multiscale formulation to Large Eddy Simulation (LES) in fluid-dynamics, using the Fourier basis.

The "senior" lecturers were complemented by four "junior" speakers, who gave account of supplementary material, detailed examples or applications. Ken Jansen, from Rensselaer Polytechnic Institute in Troy, discussed variational multiscale methods for LES using a hierarchical basis and finite elements. Joe Pasciak, from Texas A&M University, extended the multigrid and multilevel approach presented by Bramble to the relevant case of symmetric indefinite second order elliptic problems. Rob Stevenson, from Utrecht University, reported on the construction of finite element wavelets on general domains and manifolds, i.e., wavelet bases for standard finite element spaces. Karsten Urban, from RWTH in Aachen, illustrated the construction of orthogonal and biorthogonal wavelet bases in complex geometries by the domain decomposition and mapping approach.

Both the senior and the junior lecturers contributed to the scientific success of the course, which was attended by 48 participants from 13 different countries. Not only the speakers presented their own material and perspective in the most effective manner, but they also made a valuable effort to dynamically establishing cross-references with other lecturers' topics, leading to a unitary picture of the course theme.

On Tuesday, September 11, we were about to head for the afternoon session, when we were hit by the terrible news coming from New York City. Incredulity, astonishment, horror, anger, worry (particularly for the families of our American friends) were the sentiments that alternated in our hearts. No space for Mathematics was left in our minds. But on the next day, we unanimously decided to resume the course with even more determination than before; we strongly believe, and we wanted to testify, that only rationality can defeat irrationality, that only the free circulation of ideas and the mutual exchange of experiences, as it occurs in science, can defeat darkness and terror.

The present volume collects the expanded version of the lecture notes by Jim Bramble, Albert Cohen and Wolfgang Dahmen. I am grateful to them for the timely production of such high quality scientific material.

As the scientific director of the course, I wish to thank the former Director of C.I.M.E., Arrigo Cellina, and the whole Scientific Board of the Centre, for inviting me to organize the event, and for providing us the nice facilities in Martina Franca as well as part of the financial support. Special thanks are due to the Secretary of C.I.M.E., Vincenzo Vespri. Generous funding for the course was provided by the I.N.D.A.M. Groups G.N.C.S. and G.N.A.M.P.A. Support also came from the Italian Research Project M.U.R.S.T. Cofin 2000 "Calcolo Scientifico: Modelli e Metodi Numerici Innovativi" and from the European Union T.M.R. Project "Wavelets in Numerical Simulation".

The organization and the realization of the school would have been by far less successful without the superb managing skills and the generous help of Anita Tabacco. A number of logistic problems were handled and solved by Stefano Berrone, as usual in the most efficient way. The help of Dino Ricchiuti, staff member of the Dipartimento di Matematica at the Politecnico di Torino, is gratefully acknowledged. Finally, I wish to thank Giuseppe Ghibò for his accurate job of processing the electronic version of the notes.

Torino, February 2003 *Claudio Canuto*

CIME's activity is supported by:

Ministero dell'Università Ricerca Scientifica e Tecnologica COFIN '99;
Ministero degli Affari Esteri - Direzione Generale per la Promozione e la Cooperazione - Ufficio V;
Consiglio Nazionale delle Ricerche;
E.U. under the Training and Mobility of Researchers Programme;
UNESCO - ROSTE, Venice Office.

Contents

Theoretical, Applied and Computational Aspects of Nonlinear Approximation

Albert Cohen

Laboratoire d'Analyse Numérique, Université Pierre et Marie Curie, Paris
cohen@ann.jussieu.fr

Summary. Nonlinear approximation has recently found computational applications such as data compression, statistical estimation or adaptive schemes for partial differential or integral equations, especially through the development of wavelet-based methods. The goal of this paper is to provide a short survey of nonlinear approximation in the perspective of these applications, as well as to stress some remaining open areas.

1 Introduction

Approximation theory is the branch of mathematics which studies the process of approximating general functions by simple functions such as polynomials, finite elements or Fourier series. It plays therefore a central role in the accuracy analysis of numerical methods. Numerous problems of approximation theory have in common the following general setting: we are given a family of subspaces $(S_N)_{N\geq 0}$ of a normed space X, and for $f \in X$, we consider the *best approximation error*

$$\sigma_N(f) := \inf_{g\in S_N} \|f-g\|_X. \tag{1}$$

Typically, N represents the number of parameters needed to describe an element in S_N, and in most cases of interest, $\sigma_N(f)$ goes to zero as this number tends to infinity.

For a given f, we can then study the *rate of approximation*, i.e., the range of $r \geq 0$ for which there exists $C > 0$ such that

$$\sigma_N(f) \leq CN^{-r}. \tag{2}$$

Note that in order to study such an asymptotic behaviour, we can use a sequence of *near-best approximation*, i.e., $f_N \in S_N$ such that

$$\|f-f_N\|_X \leq C\sigma_N(f), \tag{3}$$

with $C > 1$ independent of N. Such a sequence always exists even when the infimum is not attained in (1), and clearly (2) is equivalent to the same estimate with $\|f - f_N\|_X$ in place of $\sigma_N(f)$.

Linear approximation deals with the situation when the S_N are linear subspaces. Classical instances of linear approximation families are the following:
1) Polynomial approximation: $S_N := \Pi_N$, the space of algebraic polynomials of degree N.
2) Spline approximation with uniform knots: some integers $0 \leq k < m$ being fixed, S_N is the spline space on $[0,1]$, consisting of C^k piecewise polynomial functions of degree m on the intervals $[j/N, (j+1)/N]$, $j = 0, \cdots, N-1$.
3) Finite element approximation on fixed triangulations: S_N are finite element spaces associated with triangulations $\mathcal{T}_N$ where N is the number of triangles in $\mathcal{T}_N$.
4) Linear approximation in a basis: given a basis $(e_k)_{k\geq 0}$ in a Banach space, $S_N := \text{Span}(e_0, \cdots, e_N)$.
In all these instances, N is typically the dimension of S_N, possibly up to some multiplicative constant.

Nonlinear approximation addresses in contrast the situation where the S_N are not linear spaces, but are still typically characterized by $\mathcal{O}(N)$ parameters. Instances of nonlinear approximation families are the following:
1) Rational approximation: $S_N := \{\frac{p}{q} \; ; \; p, q \in \Pi_N\}$, the set rational functions of degree N.
2) Free knot spline approximation: some integers $0 \leq k < m$ being fixed, S_N is the spline space on $[0,1]$ with N free knots, consisting of C^k piecewise polynomial functions of degree m on intervals $[x_j, x_{j+1}]$, for all partitions $0 = x_0 < x_1 \cdots < x_{N-1} < x_N = 1$.
3) Adaptive finite element approximation: S_N are the union of finite element spaces $V_{\mathcal{T}}$ of some fixed type associated to all triangulations $\mathcal{T}$ of cardinality less or equal to N.
4) N-term approximation in a basis: given a basis $(e_k)_{k\geq 0}$ in a Banach space, S_N is the set of all possible combinations $\sum_{k\in E} x_k e_k$ with $\#(E) \leq N$.
Note that these examples are in some sense nonlinear generalizations of the previous linear examples, since they include each of them as particular subsets. Also note that in all of these examples (except for the splines with uniform knots), we have the natural property $S_N \subset S_{N+1}$, which expresses that the approximation is "refined" as N grows.

On a theoretical level, a basic problem, both for linear and nonlinear approximation can be stated as follows:

Problem 1: *Given a nonlinear family $(S_N)_{N\geq 0}$, what are the analytic properties of a function f which ensure a prescribed rate $\sigma_N(f) \leq CN^{-r}$?*

By "analytic properties", we typically have in mind smoothness, since we know that in many contexts a prescribed rate r can be achieved provided that f belongs to some smoothness class $X_r \subset X$. Ideally, one might hope to identify the *maximal class* X_r such that the rate r is ensured, i.e., have a sharp result of the type

$$f \in X_r \Leftrightarrow \sigma_N(f) \leq CN^{-r}. \tag{4}$$

Another basic problem, perhaps on a slightly more applied level, is the effective construction of near-best approximants.

Problem 2: *Given a nonlinear family $(S_N)_{N\geq 0}$, find a simple implementable procedure $f \mapsto f_N \in S_N$ such that $\|f - f_N\|_X \leq C\sigma_N(f)$ for all $N \geq 0$.*

In the case of linear approximation, this question is usually solved if we can find a sequence of projectors $P_N : X \mapsto S_N$ such that $\|P_N\|_{X\to X} \leq K$ with K independent of N (in this case, simply take $f_N = P_N f$ and remark that $\|f - f_N\|_X \leq (1 + K)\sigma_N(f)$). It is in general a more difficult problem in the case of nonlinear method. Since the 1960's, research in approximation theory has evolved significantly toward nonlinear methods, in particular solving the two above problems for various spaces S_N.

More recently, nonlinear approximation became attractive on a more applied level, as a tool to understand and analyze the performance of *adaptive methods* in signal and image processing, statistics and numerical simulation. This is in part due to the emergence of *wavelet bases* for which simple N-term approximations (derived by thresholding the coefficients) yield in some sense optimal adaptive approximations. In such applications, the problems that arise are typically the following ones.

Problem 3 (data compression): *How can we exploit the reduction of parameters in the approximation of f by $f_N \in S_N$ in the perspective of optimally encoding f by a small number of bits ? This raises the question of a proper quantization of these parameters.*

Problem 4 (statistical estimation): *Can we use nonlinear approximation as a denoising scheme ? In this perspective, we need to understand the interplay between the approximation process and the presence of noise.*

Problem 5 (numerical simulation): *How can we compute a proper nonlinear approximation of a function u which is not given to us as a data but as the solution of some problem $F(u) = 0$? This is in particular the goal of adaptive refinement strategies in the numerical treatment of PDE's.*

The goal of the present paper is to briefly survey the subject of nonlinear approximation, with a particular focus on questions 1 to 5, and some emphasis on wavelet-based methods. We would like to point out that these questions are also addressed in the survey paper [15] which contains a more substantial development on the theoretical aspects. We hope that our notes might be helpful to the non-expert reader who wants to get a first general and intuitive vision of the subject, from the point of view of its various applications, before perhaps going into a more detailed study.

The paper is organized as follows. As a starter, we discuss in §2 a simple example, based on piecewise constant functions, which illustrate the differences between linear and nonlinear approximation, and we discuss a first algorithm which produces nonlinear piecewise constant approximations. In §3, we show that such approximations can also be produced by thresholding the coefficients in the Haar wavelet system. In §4, we give the general results on linear uniform approximation of finite element or wavelet types. General results on nonlinear adaptive approximations by wavelet thresholding or adaptive partitions are given in §5. Applications to signal compression and estimation are discussed in §6 and §7. Applications to adaptive numerical simulation are shortly described in §8. Finally, we conclude in §9 by some remarks and open problems arising naturally in the multivariate setting.

2 A Simple Example

Let us consider the approximation of functions defined on the unit interval $I = [0,1]$ by piecewise constant functions. More precisely, given a disjoint partition of I into N subintervals $I_0, \cdots, I_{N-1}$ and a function f in $L^1(I)$, we shall approximate f on each I_k by its average $a_{I_k}(f) = |I_k|^{-1} \int_{I_k} f(t)dt$. The resulting approximant can thus be written as

$$f_N := \sum_{k=1}^{N} a_{I_k}(f)\chi_{I_k}. \tag{5}$$

If the I_k are fixed independently of f, then f_N is simply the orthogonal projection of f onto the space of piecewise constant functions on the partition I_k, i.e., a *linear approximation* of f. A natural choice is the uniform partition $I_k := [k/N, (k+1)/N]$. With such a choice, let us now consider the error between f and f_N, for example in the L^∞ metric. For this, we shall assume that f is in $C(I)$, the space of continuous functions on I. It is then clear that on each I_k we have

$$|f(t) - f_N(t)| = |f(t) - a_{I_k}(f)| \leq \sup_{t,u \in I_k} |f(t) - f(u)|. \tag{6}$$

We thus have the error estimate

$$\|f - f_N\|_{L^\infty} \leq \sup_k \sup_{t,u \in I_k} |f(t) - f(u)|. \tag{7}$$

This can be converted into an estimate in terms of N, under some additional smoothness assumptions on f. In particular, if f has a bounded first derivative, we have $\sup_{t,u \in I_k} |f(t) - f(u)| \leq |I_k| \|f'\|_{L^\infty} = N^{-1} \|f'\|_{L^\infty}$, and thus

$$\|f - f_N\|_{L^\infty} \leq N^{-1}\|f'\|_{L^\infty}. \tag{8}$$

Similarly, if f is in the Hölder space C^α for some $\alpha \in]0,1[$, i.e., if for all $x, y \in [0,1]$,

$$|f(x) - f(y)| \leq C|x - y|^\alpha, \tag{9}$$

we obtain the estimate

$$\|f - f_N\|_{L^\infty} \leq CN^{-\alpha}. \tag{10}$$

By considering simple examples such as $f(x) = x^\alpha$ for $0 < \alpha \leq 1$, one can easily check that this rate is actually sharp. In fact it is an easy exercise to check that a converse result holds : if a function $f \in C([0,1])$ satisfies (10) for some $\alpha \in]0,1[$ then necessarily f is in C^α, and f' is in L^∞ in the case where $\alpha = 1$. Finally note that we cannot hope for a better rate than N^{-1}: this reflects the fact that piecewise constant functions are only first order accurate.

If we now consider an *adaptive partition* where the I_k depend on the function f itself, we enter the topic of *nonlinear approximation*. In order to understand the potential gain in switching from uniform to adaptive partitions, let us consider a function f such that f' is integrable, i.e., f is in the space $W^{1,1}$. Since we have $\sup_{t,u \in I_k} |f(t) - f(u)| \leq \int_{I_k} |f'(t)|dt$, we see that a natural choice of the I_k can be made by equalizing the quantities $\int_{I_k} |f'(t)|dt = N^{-1} \int_0^1 |f'(t)|dt$, so that, in view of the basic estimate (7), we obtain the error estimate

$$\|f - f_N\|_{L^\infty} \leq N^{-1}\|f'\|_{L^1}. \tag{11}$$

In comparison with the uniform/linear situation, we thus have obtained the same rate as in (8) for a larger class of functions, since f' is not assumed to be bounded but only integrable. On a slightly different angle, the nonlinear approximation rate might be significantly better than the linear rate for a fixed function f. For instance, the function $f(x) = x^\alpha$, $0 < \alpha \leq 1$, has the linear rate $N^{-\alpha}$ and the nonlinear rate N^{-1} since $f'(x) = \alpha x^{\alpha-1}$ is in $L^1(I)$. Similarly to the linear case, it can be checked that a converse result holds : if $f \in C([0,1])$ is such that

$$\sigma_N(f) \leq CN^{-1}, \tag{12}$$

where $\sigma_N(f)$ is the L^∞ error of best approximation by adaptive piecewise constant functions on N intervals, then f is necessarily in $W^{1,1}$.

The above construction of an adaptive partition based on balancing the L^1 norm of f' is somehow theoretical, in the sense that it pre-assumes a certain amount of smoothness for f. A more realistic adaptive approximation algorithm should also operate on functions which are not in $W^{1,1}$. We shall describe two natural algorithms for building an adaptive partition. The first algorithm is sometimes known as *adaptive splitting* and was studied e.g. in [17]. In this algorithm, the partition is determined by a prescribed tolerance $\varepsilon > 0$

which represents the accuracy that one wishes to achieve. Given a partition of $[0,1]$, and any interval I_k of this partition, we split I_k into two sub-intervals of equal size if $\|f - a_{I_k}(f)\|_{L^\infty(I_k)} \geq \varepsilon$ or leave it as such otherwise. Starting this procedure on the single $I = [0,1]$ and using a fixed tolerance $\varepsilon > 0$ at each step, we end with an adaptive partition $(I_1, \cdots, I_N)$ with $N(\varepsilon)$ and a corresponding piecewise constant approximation f_N with $N = N(\varepsilon)$ pieces such that $\|f - f_N\|_{L^\infty} \leq \varepsilon$. Note that we now have the restriction that the I_k are *dyadic intervals*, i.e., intervals of the type $2^{-j}[n, n+1]$.

We now want to understand how the adaptive splitting algorithm behaves in comparison to the optimal partition. In particular, do we also have that $\|f - f_N\|_{L^\infty} \leq CN^{-1}$ when $f' \in L^1$? The answer to this question turns out to be negative, but a slight strengthening of the smoothness assumption will be sufficient to ensure this convergence rate : we shall instead assume that the *maximal function* of f' is in L^1. We recall that the maximal function of a locally integrable function g is defined by

$$Mg(x) := \sup_{r>0} [\mathrm{vol}(B(x,r)]^{-1} \int_{B(x,r)} |g(t)|dt. \tag{13}$$

It is known that $Mg \in L^p$ if and only if $g \in L^p$ for $1 < p < \infty$ and that $Mg \in L^1$ if and only if $g \in L\log L$, i.e., $\int |g| + \int |g \log |g|| < \infty$. Therefore, the assumption that Mf' is integrable is only slightly stronger than $f \in W^{1,1}$.

If $(I_1, \cdots, I_N)$ is the final partition, consider for each k the interval J_k which is the *parent* of I_k in the splitting process, i.e., such that $I_k \subset J_k$ and $|J_k| = 2|I_K|$. We therefore have

$$\varepsilon \leq \|f - a_{J_k}(f)\|_{L^\infty} \leq \int_{J_k} |f'(t)|dt. \tag{14}$$

For all $x \in I_k$, the ball $B(x, 2|I_k|)$ contains J_k and it follows therefore that

$$Mf'(x) \geq [\mathrm{vol}(B(x, 2|I_k|)]^{-1} \int_{B(x,2|I_k|)} |f'(t)|dt \geq [4|I_k|]^{-1}\varepsilon, \tag{15}$$

which implies in turn

$$\int_{I_k} Mf'(t)dt \geq \varepsilon/4. \tag{16}$$

If Mf' is integrable, this yields the estimate

$$N(\varepsilon) \leq 4\varepsilon^{-1} \int_0^1 Mf'(t)dt. \tag{17}$$

It follows that

$$\|f - f_N\|_{L^\infty} \leq CN^{-1} \tag{18}$$

with $C = 4\int_0^1 Mf'$. Note that in this case this is only a sufficient condition for the rate N^{-1} (a simple smoothness condition which characterizes this rate is still unknown).

3 The Haar System and Thresholding

The second algorithm is based on *thresholding* the decomposition of f in the simplest wavelet basis, namely the Haar system. The decomposition of a function f defined on $[0,1]$ into the Haar system is illustrated on Figure 1. The first component in this decomposition is the average of f, i.e., the projection onto the constant function $\varphi = \chi_{[0,1]}$, i.e.,

$$P_0 f = \langle f, \varphi \rangle \varphi. \tag{19}$$

The approximation is then recursively refined into

$$P_j f = \sum_{k=0}^{2^j-1} \langle f, \varphi_{j,k} \rangle \varphi_{j,k}, \tag{20}$$

where $\varphi_{j,k} = 2^{j/2}\varphi(2^j \cdot -k)$, i.e., averages of f on the intervals $I_{j,k} = [2^{-j}k, 2^{-j}(k+1)[$, $k = 0, \cdots, 2^j - 1$. Clearly $P_j f$ is the L^2-orthogonal projection of f onto the space V_j of piecewise constant functions on the intervals $I_{j,k}$, $k = 0, \cdots, 2^j - 1$. The orthogonal complement $Q_j f = P_{j+1}f - P_j f$ is spanned by the basis functions

$$\psi_{j,k} = 2^{j/2}\psi(2^j \cdot -k), \;\; k = 0, \cdots, 2^j - 1, \tag{21}$$

where ψ is 1 on $[0, 1/2[$, -1 on $[1/2, 1[$ and 0 elsewhere. By letting j go to $+\infty$, we therefore obtain the expansion of f into an orthonormal system of $L^2([0,1])$

$$f = \langle f, \varphi \rangle \varphi + \sum_{j \geq 0} \sum_{k=0}^{2^j-1} \langle f, \psi_{j,k} \rangle \psi_{j,k} = \sum_{\lambda} d_\lambda \psi_\lambda. \tag{22}$$

Here we use the notation ψ_λ and $d_\lambda = \langle f, \psi_\lambda \rangle$ in order to concatenate the scale and space parameters j and k into one index $\lambda = (j,k)$, which varies in a suitable set ∇, and to include the very first function φ into the same notation. We shall keep track of the scale by using the notation

$$|\lambda| = j \tag{23}$$

whenever the basis function ψ_λ has resolution 2^{-j}. This simple example is known as the *Haar system* since its introduction by Haar in 1909. Its main limitation is that it is based on piecewise constant functions which are discontinuous and only allow for approximation of low order accuracy. We shall remedy to this defect by using smoother wavelet bases in the next sections.

We can use wavelets in a rather trivial way to build linear approximations of a function f since the projections of f onto V_j are given by

$$P_j f = \sum_{|\lambda|<j} \sum_{\lambda} d_\lambda \psi_\lambda. \tag{24}$$

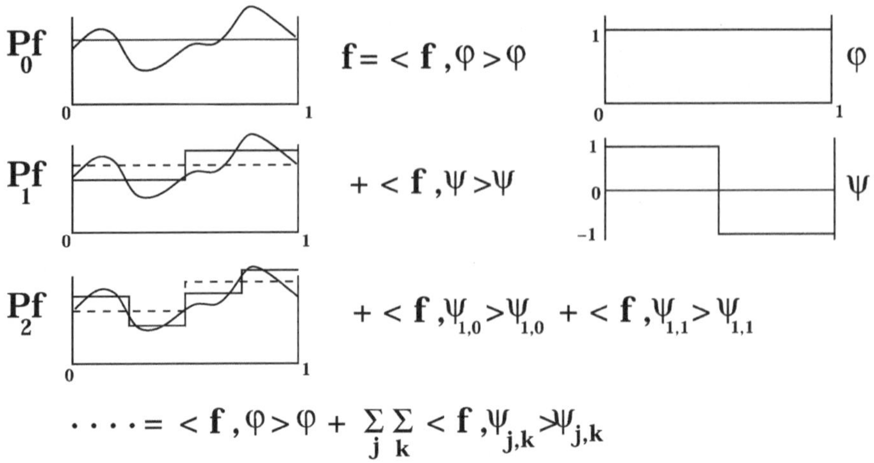

Figure 1. Decomposition into the Haar system

Such approximations simply correspond to the case $N = 2^j$ using the linear projection onto piecewise constant function on a uniform partition of N intervals, as studied in the previous section.

On the other hand, one can think of using only a restricted set of wavelet at each scale j in order to build nonlinear adaptive approximations. A natural way to obtain such adaptive approximation is by *thresholding*, i.e., keeping only the largest contributions $d_\lambda \psi_\lambda$ in the wavelet expansion of f. Such a strategy will lead to an adaptive discretization of f due to the fact that the size of wavelet coefficients d_λ is influenced by the local smoothness of f. Indeed, if f is simply bounded on the support S_λ of ψ_λ, we have the obvious estimate

$$|d_\lambda| = |\langle f, \psi_\lambda \rangle| \leq \sup_{t \in S_\lambda} |f(t)| \int |\psi_\lambda| = 2^{-|\lambda|/2} \sup_{t \in S_\lambda} |f(t)|. \qquad (25)$$

On the other hand, if f is continuously differentiable on S_λ, we can use the fact that $\int \psi_\lambda = 0$ to write

$$\begin{aligned} |d_\lambda| &= \inf_{c \in \mathbb{R}} |\langle f - c, \psi_\lambda \rangle| \\ &\leq 2^{-|\lambda|/2} \inf_{c \in \mathbb{R}} \sup_{t \in S_\lambda} |f(t) - c| \\ &\leq 2^{-|\lambda|/2} \sup_{t,u \in S_\lambda} |f(t) - f(u)| \\ &\leq 2^{-3|\lambda|/2} \sup_{t \in S_\lambda} |f'(t)|. \end{aligned}$$

Note that if f were not differentiable on S_λ but simply Hölder continuous of exponent $\alpha \in]0, 1[$, a similar computation would yield the intermediate estimate $|d_\lambda| \leq C 2^{-(\alpha+1/2)|\lambda|}$. As in the case of Fourier coefficients, more smoothness implies a faster decay, yet a fundamental difference is that only *local smoothness* is involved in the wavelet estimates. Therefore, if f is C^1 everywhere except at some isolated point x , the estimation of $|d_\lambda|$ by $2^{-3|\lambda|/2}$ will only be lost for those λ such that $x \in S_\lambda$. In that sense, multiscale representations are better adapted than Fourier representations to concentrate the information contained in functions which are not uniformly smooth.

This is illustrated by the following example. We display on Figure 2 the function $f(x) = \sqrt{|\cos(2\pi x)|}$, which has a cusp singularity at points $x = 1/4$ and $x = 3/4$, and which is discretized at resolution 2^{-13}, in order to compute its coefficients in the Haar basis for $|\lambda| < 13$. In order to visualize the effect of local smoothness on these coefficients, we display on Figure 3 the set of indices $\lambda = (j, k)$ such that $|d_\lambda|$ is larger than the threshold $\varepsilon = 5 \times 10^{-3}$, measuring the spatial position of the wavelet $2^{-j}k$ in the x axis and its scale level j in the y axis. We observe that for $j > 4$, the coefficients above the threshold are only concentrated in the vicinity of the singularities. This is explained by the fact that the decay of the coefficients is governed by $|d_\lambda| \leq 2^{-3|\lambda|/2} \sup_{t \in S_\lambda} |f'(t)|$ in the regions of smoothness, while the estimate $|d_\lambda| \leq C2^{-(\alpha+1/2)|\lambda|}$ with $\alpha = 1/2$ will prevail near the singularities. Figure 4 displays the result of the reconstruction of f using only this restricted set of wavelet coefficients,

$$f_\varepsilon = \sum_{|d_\lambda| > \varepsilon} d_\lambda \psi_\lambda, \tag{26}$$

and it reveals the *spatial adaptivity* of the thresholding operator: the approximation is automatically refined in the neighbourhood of the singularities where wavelet coefficients have been kept up to the resolution level $j = 8$. In this example, we have kept the largest components $d_\lambda \psi_\lambda$ measured in the L^2 norm. This strategy is ideal to minimize the L^2 error of approximation for a prescribed number N of preserved coefficients. If we are interested in the L^∞ error, we shall rather choose to keep the largest components measured in the L^∞ norm, i.e., the largest normalized coefficients $|d_\lambda| 2^{|\lambda|/2}$.

Just as in the case of the adaptive splitting algorithm, we might want to understand how the partition obtained by wavelet thresholding behaves in comparison to the optimal partition. The answer is again that it is nearly optimal, however we leave this question aside since we shall provide much more general results on the performance of wavelet thresholding in §4. The wavelet approach to nonlinear approximation is particularly attractive for the following reason: in this approach, the nonlinearity is reduced to a very simple operation (thresholding according to the size of the coefficients), resulting in simple and efficient algorithms for dealing with many applications, as well as a relatively simple analysis of these applications.

4 Linear Uniform Approximation

We now address linear uniform approximation in more general terms. In order to improve on the rate N^{-1} obtained with piecewise constant functions, one needs to introduce approximants with a higher degree of accuracy, such as splines or finite element spaces. In the case of linear uniform approximation,

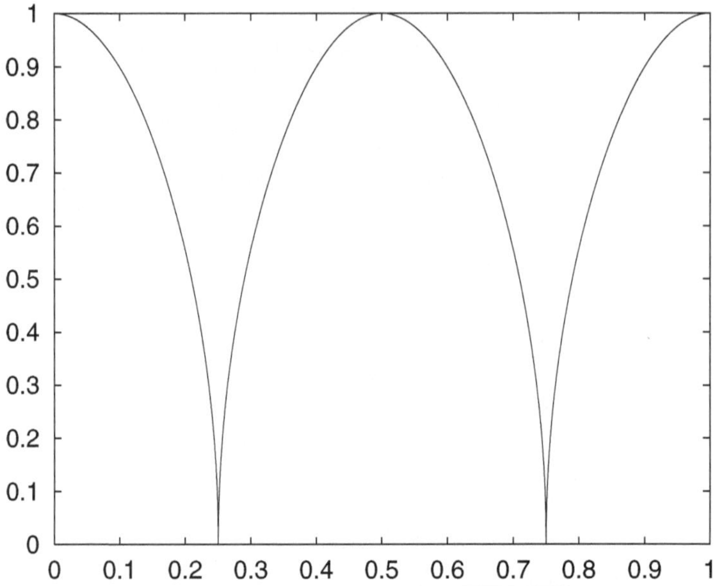

Figure 2. The function $f(x) = \sqrt{|\cos(2\pi x)|}$

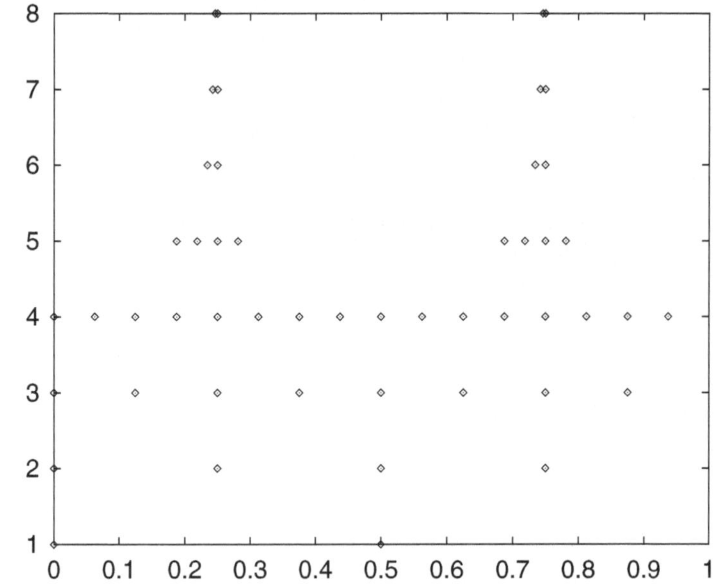

Figure 3. Coefficients above threshold $\varepsilon = 5 \times 10^{-3}$

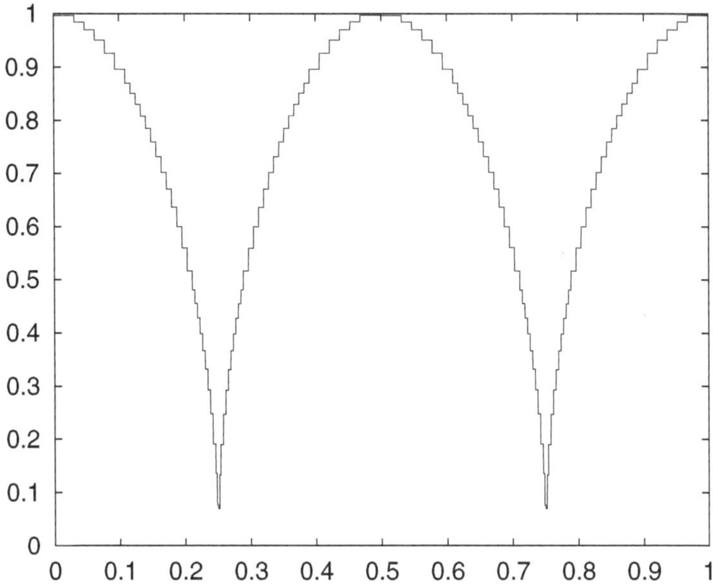

Figure 4. Reconstruction from coefficients above threshold

these spaces consists of piecewise polynomial functions onto regular partitions $\mathcal{T}_h$ with uniform mesh size h. If V_h is such a space discretizing a regular domain $\Omega \subset \mathbb{R}^d$, its dimension is therefore of the same order as the number of balls of radius h which are needed to cover Ω, namely

$$N = \dim(V_h) \sim h^{-d}. \quad (27)$$

The approximation theory for such spaces is quite classical, see, e.g., [5], and can be summarized in the following way. If $W^{s,p}$ denotes the classical Sobolev space, consisting of those functions in L^p such that $D^\alpha f \in L^p$ for $|\alpha| \leq s$, we typically have the error estimate

$$\inf_{g \in V_h} \|f - g\|_{W^{s,p}} \leq Ch^t \|f\|_{W^{s+t,p}} \quad (28)$$

provided that V_h is contained in $W^{s,p}$ and that V_h has approximation order larger than $s+t$, i.e., contains all polynomials of degree strictly less than $s+t$. In the particular case $s = 0$, this gives

$$\inf_{g \in V_h} \|f - g\|_{L^p} \leq Ch^t \|f\|_{W^{t,p}}. \quad (29)$$

Such classical results also hold for fractional smoothness. If we rewrite them in terms of the decay of the best approximation error with respect to the number of parameters, we therefore obtain that if $X = W^{s,p}$, we have

$$\sigma_N(f) \leq CN^{-t/d}, \quad (30)$$

provided that f has t additional derivatives in the metric L^p compared to the general functions in X. Therefore, the compromise between the L^p or $W^{s,p}$ approximation error and the number of parameters is governed by the approximation order of the V_h spaces, the dimension d and the level of smoothness of f measured in L^p. Such approximation results can be understood at a very basic and intuitive level: if V_h contains polynomials of degree $t-1$, we can think of the approximation of f as a close substitute to its Taylor expansion f_K at this order on each element $K \in \mathcal{T}_h$, which has accuracy $h^t|D^t f|$, and (29) can then be thought as the integrated version of this local error estimate.

At this stage it is interesting to look at linear approximation from the angle of multiscale decompositions into wavelet bases. Such bases are generalizations of the Haar system which was discussed in the previous section, and we shall first recall their main features (see [14] and [6] for more details). They are associated with multiresolution approximation spaces $(V_j)_{j\geq 0}$ such that $V_j \subset V_{j+1}$ and V_j is generated by a local basis $(\varphi_\lambda)_{|\lambda|=j}$. By *local* we mean that the supports are controlled by

$$\mathrm{diam}(\mathrm{supp}(\varphi_\lambda)) \leq C 2^{-j} \tag{31}$$

if $\lambda \in \Gamma_j$ and are "almost disjoint" in the sense that

$$\#\{\mu \in \Gamma_j \text{ s.t. } \mathrm{supp}(\varphi_\lambda) \cap \mathrm{supp}(\varphi_\mu) \neq \emptyset\} \leq C \tag{32}$$

with C independent of λ and j. Such spaces can be built in particular as nested finite element spaces $V_j = V_{h_j}$ with mesh size $h_j \approx 2^{-j}$ in which case $(\varphi_\lambda)_{|\lambda|=j}$ is the classical nodal basis. Just as in the case of the Haar system, a complement space W_j of V_j into V_{j+1} is generated by a similar local basis $(\psi_\lambda)_{|\lambda|=j}$. The full multiscale wavelet basis (ψ_λ), allows us to expand an arbitrary function f with the convention that we incorporate the functions $(\varphi_\lambda)_{|\lambda|=0}$ into the first "layer" $(\psi_\lambda)_{|\lambda|=0}$. In the standard constructions of wavelets on the euclidean space $\mathbb{R}^d$, the scaling functions have the form $\varphi_\lambda = \varphi_{j,k} = 2^{jd/2}\varphi(2^j\cdot - k)$, $k \in \mathbb{Z}^d$ and similarly for the wavelets, so that $\lambda = (j,k)$. In the case of a general domain $\Omega \in \mathbb{R}^d$, special adaptations of the basis functions are required near the boundary $\partial\Omega$, which are accounted in the general notations λ. Wavelets need not be orthonormal, but one often requires that they constitute a Riesz basis of $L^2(\Omega)$, i.e., their finite linear combinations are dense in L^2 and for all sequences (d_λ) we have the norm equivalence

$$\|\sum_\lambda d_\lambda \psi_\lambda\|_{L^2}^2 \sim \sum_\lambda |d_\lambda|^2. \tag{33}$$

In such a case, the coefficients d_λ in the expansion of f are obtained by an inner product $d_\lambda = \langle f, \tilde{\psi}_\lambda \rangle$, where the dual wavelet $\tilde{\psi}_\lambda$ is an L^2-function. In the standard biorthogonal constructions, the dual wavelet system $(\tilde{\psi}_\lambda)$ is also built from nested spaces $\tilde{V}_j$ and has similar local support properties as the primal wavelets ψ_λ. The practical advantage of such a setting is the possibility of "switching" between the "standard" (or "nodal") discretization of $f \in V_j$ in the basis $(\varphi_\lambda)_{|\lambda|=j}$ and its "multiscale" representation in the basis $(\psi_\lambda)_{|\lambda|<j}$ by means of fast $\mathcal{O}(N)$ decomposition and reconstruction algorithms, where $N \sim 2^{dj}$ denotes the dimension of V_j in the case where Ω is bounded.

Multiscale approximations and decompositions into wavelet bases will provide a slightly stronger statement of the linear approximation results, due to the possibility of characterizing the smoothness of a function f through the numerical properties of its multiscale decomposition. In the case of Sobolev spaces $H^s = W^{s,2}$, this characterization has the form of the following norm equivalence

$$\|f\|_{H^s}^2 \sim \|P_0 f\|_{L^2}^2 + \sum_{j\geq 0} 2^{2sj} \|Q_j f\|_{L^2}^2, \tag{34}$$

where $P_j f = \sum_{|\lambda|<j} d_\lambda \psi_\lambda$ and $Q_j f = P_{j+1} f - P_j f = \sum_{|\lambda|<j} d_\lambda \psi_\lambda$ are respectively the biorthogonal projectors onto V_j and W_j. Such a norm equivalence should at least be understood at the intuitive level as a close substitute to the Fourier characterization

$$\|f\|_{H^s}^2 \sim \int (1+|\omega|^{2s})|\hat{f}(\omega)|^2 d\omega, \tag{35}$$

where the weight $(1+|\omega|^{2s})$ plays an analogous role as 2^{2sj} in (34). Note that in the particular case where V_j is the space of functions such that $\hat{f}$ is supported in $[-2^j, 2^j]$ and P_j the orthogonal projector, we directly obtain

$$\begin{aligned}
\|f\|_{H^s}^2 &\sim \textstyle\int (1+|\omega|^{2s})|\hat{f}(\omega)|^2 d\omega \\
&= \textstyle\int_{|\omega|<1} (1+|\omega|^{2s})|\hat{f}(\omega)|^2 + \sum_{j\geq 0} \int_{2^j<|\omega|<2^{j+1}} (1+|\omega|^{2s})|\hat{f}(\omega)|^2 \\
&\sim \textstyle\int_{|\omega|<1} |\hat{f}(\omega)|^2 + \sum_{j\geq 0} 2^{2sj} \int_{2^j<|\omega|<2^{j+1}} |\hat{f}(\omega)|^2 \\
&\sim \|P_0 f\|_{L^2}^2 + \textstyle\sum_{j\geq 0} 2^{2sj} \|Q_j f\|_{L^2}^2.
\end{aligned}$$

We next remark that Q_j can be replaced by $I - P_j$ in (34), in the sense that we also have

$$\|f\|_{H^s}^2 \sim \|P_0 f\|_{L^2}^2 + \sum_{j\geq 0} 2^{2sj} \|f - P_j f\|_{L^2}^2. \tag{36}$$

In order to prove this, we need to compare the right hand side of (34) and (36). In one direction we obviously have

$$\|Q_j f\| \leq \|f - P_j f\| + \|f - P_{j+1} f\| \tag{37}$$

which is sufficient to control the r.h.s. of (34) by the r.h.s. of (36). In the other direction we use

$$\|f - P_j f\| \leq \sum_{l\geq j} \|Q_l f\| \tag{38}$$

and conclude by the discrete Hardy inequality which states that if (a_j) is a positive sequence and $b_j := \sum_{l\geq j} a_j$, then for all $s > 0$ and $p > 0$

$$\|(2^{sj}b_j)\|_{\ell^p} \leq C(s,p)\|(2^{sj}a_j)\|_{\ell^p}. \tag{39}$$

The norm equivalence (36) show that $f \in H^s$ is exactly equivalent to the property that

$$(2^{sj} \inf_{g \in V_j} \|f - g\|_{L^2})_{j \geq 0} \in \ell^2, \tag{40}$$

which is a slight improvement over the approximation rate

$$\inf_{g \in V_j} \|f - g\|_{L^2} \leq C 2^{-sj}, \tag{41}$$

which would be a re-expression of (29). In order to provide a similar statement for more general L^p approximation, one needs to introduce the Besov spaces $B^s_{p,q}$ which measure smoothness of order $s > 0$ in L^p according to

$$\|f\|_{B^s_{p,q}} := \|f\|_{L^p} + \|(2^{sj}\omega_m(f, 2^{-j})_p)_{j \geq 0}\|_{\ell^q}, \tag{42}$$

where

$$\omega_m(f,t)_p := \sup_{|h| \leq t} \|\Delta_h^m f\|_{L^p} = \sup_{|h| \leq t} \|\sum_{k=0}^{m} \binom{m}{k} (-1)^k f(\cdot - kh)\|_{L^p}$$

is the m-th order L^p modulus of smoothness and m is any integer strictly larger than s. Recall that we have $H^s \sim B^s_{2,2}$ for all $s > 0$, $C^s \sim B^s_{\infty,\infty}$ and $W^{s,p} \sim B^s_{p,p}$ for all non-integer $s > 0$ and $p \neq 2$. For such classes, the norm equivalences which generalize (34) and (36) have the form

$$\begin{aligned} \|f\|_{B^s_{p,q}} &\sim \|P_0 f\|_{L^p} + \|(2^{sj}\|Q_j f\|_{L^p})_{j \geq 0}\|_{\ell^q} \\ &\sim \|P_0 f\|_{L^p} + \|(2^{sj}\|f - P_j f\|_{L^p})_{j \geq 0}\|_{\ell^q}. \end{aligned} \tag{43}$$

Such norm equivalences reflect the intuitive idea that the linear approximation error $\|f - P_j f\|_{L^p}$ decays like $\mathcal{O}(2^{-sj})$ or $\mathcal{O}(N^{-s/d})$ provided that f has "s derivatives in L^p". Their are essentially valid under the restriction that the wavelet ψ_λ itself has slightly more than s derivative in L^p. We refer to [6] for the general mechanism which allows us to prove these results, based on direct and inverse estimates as well as interpolation theory.

Finally, we can re-express these norm equivalences in terms of wavelet coefficients: using the local properties of wavelet bases, we have at each level the norm equivalence

$$\|Q_j f\|^p_{L^p} \sim \sum_{|\lambda|=j} \|d_\lambda \psi_\lambda\|^p_{L^p} \sim \sum_{|\lambda|=j} 2^{d(p/2-1)|\lambda|}|d_\lambda|^p, \tag{44}$$

which in combination with (43) yields

$$\|f\|_{B^s_{p,q}} \sim \|(2^{(s+d/2-d/p)j}\|(d_\lambda)_{|\lambda|=j}\|_{\ell^p})_{j \geq 0}\|_{\ell^q}. \tag{45}$$

5 Nonlinear Adaptive Approximation

Let us now turn to nonlinear adaptive approximation, with a special focus on N-term approximation in a wavelet basis: denoting by

$$S_N := \{\sum_{\lambda\in E} c_\lambda\psi_\lambda \; ; \; \#(E) \leq N\}, \tag{46}$$

the set of all possible N-term combinations of wavelets, we are interested in the behaviour of $\sigma_N(f)$ as defined in (1) for some given error norm X. We first consider the case $X = L^2$ and assume for simplicity that the ψ_λ constitute an orthonormal basis. In this case, it is a straightforward computation that the best N-term approximation of a function f is achieved by its truncated expansion

$$f_N := \sum_{\lambda\in E_N(f)} d_\lambda\psi_\lambda, \tag{47}$$

where $E_N(f)$ contains the indices corresponding to the N largest $|f_\lambda|$. The approximation error is thus given by

$$\sigma_N(f) = \|f - f_N\|_{L^2} = (\sum_{\lambda\notin E_N(f)} |d_\lambda|^2)^{1/2} = (\sum_{n\geq N} d_n^2)^{1/2}, \tag{48}$$

where $(d_n)_{n\geq 0}$ is defined as the *decreasing rearrangement* of the $|d_\lambda|$, $\lambda\in\nabla$ (i.e., d_{n-1} is the n-th largest $|d_\lambda|$).

Consider now the Besov spaces $B^s_{\tau,\tau}$ where $s > 0$ and τ are linked by $1/\tau = 1/2 + s/d$. According to the norm equivalence (45) we note that these space are simply characterized by

$$\|f\|_{B^s_{\tau,\tau}} \sim \|(d_\lambda)_{\lambda\in\nabla}\|_{\ell^\tau} \tag{49}$$

Thus if $f\in B^s_{\tau,\tau}$, we find that the decreasing rearrangement $(d_n)_{n\geq 0}$ satisfies

$$nd_n^\tau \leq \sum_{k=0}^{n-1} d_k^\tau \leq \sum_{k\geq 0} d_k^\tau = \sum_{\lambda\in\nabla} d_\lambda^\tau \leq C\|f\|^\tau_{B^s_{\tau,\tau}} < +\infty, \tag{50}$$

and therefore

$$d_n \leq Cn^{-1/\tau}\|f\|_{B^s_{\tau,\tau}}. \tag{51}$$

It follows that the approximation error is bounded by

$$\sigma_N(f) \leq C\|f\|_{B^s_{\tau,\tau}}(\sum_{n\geq N} n^{-\frac{2}{\tau}})^{1/2} \leq CN^{\frac{1}{2}-\frac{1}{\tau}}\|f\|_{B^s_{\tau,\tau}} = CN^{-s/d}\|f\|_{B^s_{\tau,\tau}}. \tag{52}$$

At this stage let us make some remarks:

- As it was previously noticed, the rate $N^{-s/d}$ can be achieved by linear approximation for functions having s derivative in L^2, i.e., functions in H^s. Just as in the simple example of §2, the gain in switching to nonlinear approximation is in that the class $B^s_{\tau,\tau}$ is larger than H^s. In particular $B^s_{\tau,\tau}$ contains discontinuous functions for arbitrarily large values of s while functions in H^s are necessarily continuous if $s > d/2$.
- The rate (52) is implied by $f \in B^s_{\tau,\tau}$. On the other hand it is easy to check that (52) is equivalent to the property (51), which is itself equivalent to the property that the sequence $(d_\lambda)_{\lambda\in\nabla}$ is in the weak space ℓ^τ_w, i.e.,

$$\#\{\lambda \text{ s.t. } |d_\lambda| \geq \varepsilon\} \leq C\varepsilon^{-\tau}. \tag{53}$$

  This shows that the property $f \in B^s_{\tau,\tau}$ is almost equivalent to the rate (52). One can easily check that the exact characterization of $B^s_{\tau,\tau}$ is by the stronger property $\sum_{N\geq 0}(N^{s/d}\sigma_N(f))^\tau N^{-1} < +\infty$.
- The space $B^s_{\tau,\tau}$ is *critically embedded in* L^2 in the sense that the injection is not compact. This can be viewed as an instance of the Sobolev embedding theorem, or directly checked in terms of the non-compact embedding of ℓ^τ into ℓ^2 when $\tau \leq 2$. In particular $B^s_{\tau,\tau}$ is not contained in any Sobolev space H^s for $s > 0$. Therefore, no convergence rate can be expected for linear approximation of functions in $B^s_{\tau,\tau}$.

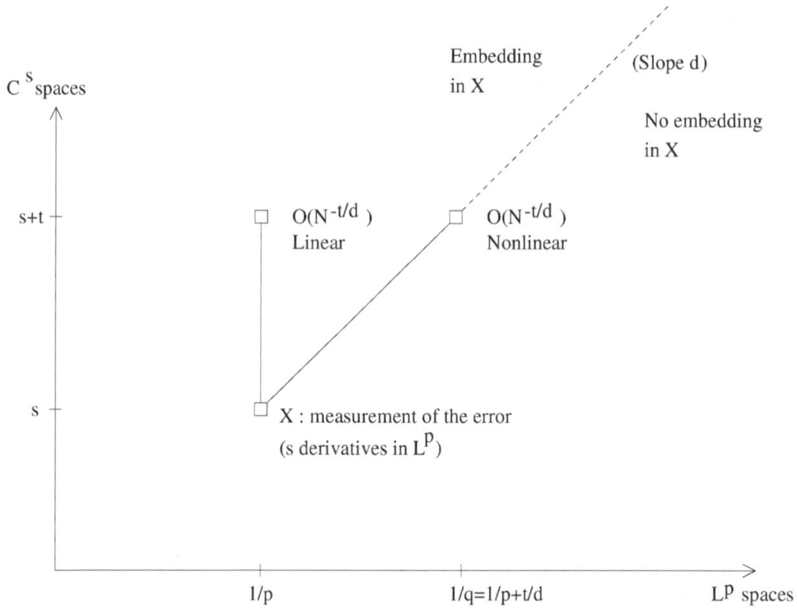

Figure 5. Pictorial interpretation of nonlinear vs linear approximation

The general theory of nonlinear wavelet approximation developed by DeVore and its collaborators extends these results to various error norms, for which the analysis is far more difficult than for the L^2 norm. This theory is fully detailed in [15], and we would like to summarize it by stressing on three main types of results, the two first answering respectively to problems 1 and 2 described in the introduction.

Approximation and smoothness spaces. Given an error norm $\|\cdot\|_X$ corresponding to some smoothness space in d-dimension, the space Y of those functions such that $\sigma_N(f) = \text{dist}_X(f, S_N) \le CN^{-t/d}$ has a typical description in terms of another smoothness space. Typically, if X represents s order of smoothness in L^p, Y will represent $s+t$ order of smoothness in L^τ with $1/\tau = 1/p + t/d$ and its injection in X is not compact. This generic result has a graphical interpretation displayed on Figure 5. On this figure, a point $(s, 1/p)$ represents function spaces with smoothness r in L^p, and the point Y sits s level of smoothness above X on the critical embedding line of slope d emanating from X. Of course in order to obtain rigorous results, one needs to specify for each case the exact meaning of "s derivative in L^p" and/or slightly modify the property $\sigma_N(f) \le CN^{-t/d}$. For instance, if $X = L^p$ for some $p \in]1, \infty[$, then $f \in B^s_{\tau,\tau} = Y$ with $1/\tau = 1/p + t/d$ if and only if $\sum_{N\ge 0}[N^{t/d}\sigma_N(f)]^\tau N^{-1} < +\infty$. One also needs to assume that the wavelet basis has enough smoothness, since it should at least be contained in Y.

Realization of a near-best approximation. For various error metric X, a near-best approximation of f in S_N is achieved by $f_N := \sum_{\lambda\in\Lambda_N(f)} d_\lambda \psi_\lambda$ where $d_\lambda := \langle f, \tilde{\psi}_\lambda\rangle$ are the wavelet coefficients of f and $\Lambda_N(f)$ is the set of indices corresponding to the N largest contributions $\|d_\lambda\psi_\lambda\|_X$. This fact is rather easy to prove when X is itself a Besov space, by using (45). A much more elaborate result is that it is also true for spaces such as L^p and $W^{m,p}$ for $1 < p < +\infty$, and for the Hardy spaces H^p when $p \le 1$ (see [21]).

Connections with other types of nonlinear approximation. In the univariate setting, the smoothness spaces Y characterized by a certain rate of nonlinear approximation in X are essentially the same if we replace N-term combinations of wavelets by splines with N free knots or by rational functions of degree N. The similarity between wavelets and free knot splines is intuitive since both methods allow the same kind of adaptive refinement, either by inserting knots or by adding wavelet components at finer scales. The similarities between free knot splines and rational approximation were elucidated by Petrushev in [19]. However, the equivalence between wavelets and these other types of approximation is no longer valid in the multivariate context (see §7). Also closely related to N-term approximations are *adaptive splitting procedures*, which are generalizations of the splitting procedure proposed in §2 to higher order piecewise polynomial approximation (see e.g. [17] and [15]). Such procedures typically aim at equilibrating the local error $\|f - f_N\|_{L^p}$ on each element of the adaptive partition. In the case of the example of §2, we remark that the piecewise constant approximation resulting from the adaptive splitting procedure can always be viewed as an N-term approximation in the Haar system, in which the involved coefficients have a certain *tree structure*: if $\lambda = (j,k)$ is used in the approximation, then $(j-1,[k/2])$ is also used at the previous coarser level. Therefore the performances of adaptive splitting approximation is essentially equivalent to those of N-term approximation with the additional tree structure restriction. These performances have been studied in [10] where it is shown that the tree structure restriction does not affect the order $N^{-s/d}$ of N-term approximation in $X \sim (1/p, r)$ if the space $Y \sim (1/\tau, r+s)$ is replaced by $\tilde{Y} \sim (1/\tilde{\tau}, r+s)$ with $1/\tilde{\tau} < 1/\tau = 1/p + s/d$.

6 Data Compression

There exist many interesting applications of wavelets to signal processing and we refer to [18] for a detailed overview. In this section and in the following one, we would like to discuss two applications which exploit the fact that certain signals - in particular images - have a sparse representation into wavelet bases. Nonlinear approximation theory allows us to "quantify" the level of sparsity in terms of the decay of the error of N-terms approximation.

On a mathematical point of view, the N-term approximation of a signal f can already be viewed as a "compression" algorithm since we are reducing the number of degrees of freedom which represent f. However, practical compression means that the approximation of f is represented by a *finite number of bits*. Wavelet-based compression algorithms are a particular case of transform coding algorithms which have the following general structure:

- Transformation: the original signal f is transformed into its representation $\mathbf{d}$ (in our case of interest, the wavelet coefficients $\mathbf{d} = (d_\lambda)$) by an invertible transform $\mathcal{R}$.
- Quantization: the representation $\mathbf{d}$ is replaced by an approximation $\tilde{\mathbf{d}}$ which can only take a finite number of values. This approximation can be encoded with a finite number of bits.
- Reconstruction: from the encoded signal, one can reconstruct $\tilde{\mathbf{d}}$ and therefore an approximation $\tilde{f} = \mathcal{R}^{-1}\tilde{\mathbf{d}}$ of the original signal f.

Therefore, a key issue is the development of appropriate quantization strategies for the wavelet representation and the analysis of the error produced by quantizing the wavelet coefficients. Such strategies should in some sense minimize the distorsion $\|f - \tilde{f}\|_X$ for a prescribed number of bits N and error metric X. Of course this program only makes sense if we refer to a certain modelization of the signal: in a deterministic context, one considers the error $\sup_{f \in Y} \|f - \tilde{f}\|_X$ for a given class Y, while in a stochastic context, one considers the error $E(\|f - \tilde{f}\|_X)$ where the expectation is over the realizations f of a stochastic process. In the following we shall indicate some results in the deterministic context.

We shall discuss here the simple case of *scalar quantization* which amounts to quantizing independently the coefficients d_λ into approximations $\tilde{d}_\lambda$ in order to produce $\tilde{\mathbf{d}}$. Similarly to the distinction between linear and nonlinear approximation, we can distinguish between two types of quantization strategies:

- Non-adaptive quantization: the map $d_\lambda \mapsto \tilde{d}_\lambda$ and the number of bits which is used to represent d_λ depend only on the index λ. In practice they typically depend on the scale level $|\lambda|$: less bits are allocated to the fine scale coefficients which have smaller values than the coarse scale coefficients *in an averaged sense.*
- Adaptive quantization: the map $d_\lambda \mapsto \tilde{d}_\lambda$ and the number of bits which is used to represent d_λ depend both of λ and of the amplitude value $|d_\lambda|$. In practice they typically depend on $|d_\lambda|$ only: more bits are allocated to the large coefficients which correspond to different indices from one signal to another.

The second strategy is clearly more appropriate in order to exploit the sparsity of the wavelet representation, since a large number of bits will be used only for a small number of numerically significant coefficients. In order to analyze this idea more precisely, let us consider the following specific strategy: for a fixed $\varepsilon > 0$, we affect no bits to the details such that $|d_\lambda| \leq \varepsilon$ by setting $\tilde{d}_\lambda = 0$, which amount in thresholding them, and we affect j bits to a detail such that $2^{j-1}\varepsilon < |d_\lambda| \leq 2^j\varepsilon$. By choosing the 2^j values of $\tilde{d}_\lambda$ uniformly in the range $]-2^j\varepsilon, -2^{j-1}\varepsilon[\cup]2^{j-1}\varepsilon, 2^j\varepsilon[$, we thus ensure that for all λ

$$|d_\lambda - \tilde{d}_\lambda| \leq \varepsilon. \tag{54}$$

If we measure the error in $X = L^2$, assuming the $\{\psi_\lambda\}$ to form a Riesz basis, we find that

$$\|f - \tilde{f}\|^2 = \sum_{\lambda\in\nabla} |d_\lambda - \tilde{d}_\lambda|^2 \leq \varepsilon^2 \#\{\lambda \ s.t. \ |d_\lambda| \geq \varepsilon\} + \sum_{|d_\lambda|\leq\varepsilon} |d_\lambda|^2. \tag{55}$$

Note that the second term is simply the error of nonlinear approximation by thresholding at the level ε, while the first term corresponds to the effect of quantizing the significant coefficients.

Let us now assume that the class of signals Y has a sparse wavelet representation in the sense that there exists $\tau \leq 2$ and $C > 0$ such that for all $f \in Y$ we have $\mathbf{d} = (d_\lambda)_{\lambda\in\nabla} \in \ell^\tau_w(\nabla)$, with $\|\mathbf{d}\|_{\ell^\tau_w} \leq C$, i.e.,

$$\sup_{f\in Y} \#\{\lambda \text{ s.t. } |d_\lambda| \geq \eta\} \leq C\eta^{-\tau}. \tag{56}$$

We have seen in the previous section that this property is satisfied when $\|f\|_{B^s_{\tau,\tau}} \leq C$ for all $f \in Y$ with $1/\tau = 1/2 + s/d$ and that it is equivalent to the nonlinear approximation property $\sigma_N \leq CN^{-s/d}$. Using (56), we can estimate both terms in (55) as follows: for the quantization term, we simply obtain

$$\varepsilon^2 \#\{\lambda \ s.t. \ |d_\lambda| \geq \varepsilon\} \leq C\varepsilon^{2-\tau}, \tag{57}$$

while for the thresholding term we have

$$\sum_{|d_\lambda| \leq \varepsilon} |d_\lambda|^2 \leq \sum_{j \geq 0} 2^{-2j} \varepsilon^2 \#\{\lambda \ s.t. \ |d_\lambda| \geq \varepsilon 2^{-j-1}\} \leq C\varepsilon^{2-\tau}. \tag{58}$$

Therefore we find that the compression error is estimated by $C\varepsilon^{1-\tau/2}$. We can also estimate the number of bits N_q which are used to quantize the d_λ according to

$$N_q = \sum_{j>0} j \#\{\lambda \text{ s.t. } 2^{j-1}\varepsilon < |d_\lambda| \leq 2^j \varepsilon\} \leq C\varepsilon^{-\tau} \sum_{j>0} j 2^{-\tau j} \leq C\varepsilon^{-\tau}. \tag{59}$$

Comparing N_q and the compression error, we find the striking result that

$$\|f - \tilde{f}\|_{L^2} \leq C N_q^{(1-\tau/2)/\tau} = C N_q^{-s/d}. \tag{60}$$

At the first sight, it seems that we obtain with only N bits the same rate as for nonlinear approximation which requires N real coefficients. However, a specific additional difficulty of adaptive quantization is that we also need to encode the *addresses* λ such that $2^{j-1}\varepsilon < |d_\lambda| \leq 2^j\varepsilon$. The bit cost N_a of this addressing can be significantly close to N_q or even higher. If the class of signals is modelized by (56), we actually find that N_a is infinite since the large coefficients could be located anywhere. In order to have $N_a \leq C\varepsilon^{-\tau}$ as well, and thus obtain the desired estimate $\|f - \tilde{f}\|_{L^2} \leq CN^{-s/d}$ with $N = N_q + N_a$, it is necessary to make some little additional assumption on Y that restricts the location of the large coefficients and to develop a suitable addressing strategy. The most efficient wavelet-compression algorithms, such as the one introduced in [20] (and further developed in the compression standard JPEG 2000), typically apply addressing strategies based on *tree structures* within the indices λ. We also refer to [10] where it is proved that such strategy allow us to recover optimal rate/distorsion bounds – i.e., optimal behaviours of the compression error with respect to the number of bits N – for various deterministic classes Y modelizing the signals.

In practice such results can only be observed for a certain range of N, since the original itself is most often given by a finite number of bits N_o, e.g. a digital image. Therefore modelizing the signal by a function class and deriving rate/distorsion bounds from this modelization is usually relevant only for low bit rate $N << N_o$, i.e., high compression ratio. One should then of course address the questions of "what are the natural deterministic classes which model real signals" and "what can one say about the sparsity of wavelet representations for these classes". An interesting example is given by real images which are often modelized by the space BV of functions with bounded variation. This function space represents functions which have one order of smoothness in L^1 in the sense that their gradient is a finite measure. This includes in particular functions of the type χ_Ω for domains Ω with boundaries of finite length. In [11] it is proved that the wavelet coefficients of a function $f \in BV$ are sparse in the sense that they are in ℓ^1_w. This allows us to expect a nonlinear approximation error in $N^{-1/2}$ for images, and a similar rate for compression provided that we can handle the addressing with a reasonable number of bits. The last task turns out to be feasible, thanks to some additional properties, such as the L^∞-boundedness of images.

7 Statistical Estimation

In recent years, wavelet-based thresholding methods have been widely applied to a large range of problems in statistics - density estimation, white noise removal, nonparametric regression, diffusion estimation - since the pioneering work of Donoho, Johnstone, Kerkyacharian and Picard (see e.g. [16]). In some sense the growing interest for thresholding strategies represent a significant "switch" from linear to nonlinear/adaptive methods. Here we shall consider the simple white noise model, i.e., given a function $f(t)$ we observe on $[0, 1]$

$$dg(t) = f(t)dt + \varepsilon dw(t), \tag{61}$$

where $w(t)$ is a Brownian motion. In other words, we observe the function f with an additive white gaussian noise of variance ε^2. This model can of course be generalized to higher dimension. We are now interested in constructing an estimator $\tilde{f}$ from the data g. The most common measure of the estimation error is in the mean square sense: assuming that $f \in L^2$ we are interested in the quantity $E(\|\tilde{f} - f\|^2_{L^2})$. Similarly to data compression, the design of an optimal estimation procedure in order to minimize the mean square error is relative to a specific modelization of the signal f either by a deterministic class Y or by a stochastic process.

Linear estimation methods define $\hat{f}$ by applying a linear operator to g. In many practical situations this operator is translation invariant and amounts to a filtering procedure, i.e., $\tilde{f} = h * g$. For example, in the case of a second order stationary process, the *Wiener filter* gives an optimal solution in terms of $\hat{h}(\omega) := \hat{r}(\omega)/(\hat{r}(\omega) + \varepsilon^2)$ where $\hat{r}(\omega)$ is the power spectrum of f, i.e., the Fourier transform of $r(u) := E(f(t)f(t + u))$. Another frequently used linear method is by projection on some finite dimensional subspace V, i.e., $\tilde{f} = Pg = \sum_{n=0}^{N} \langle g, \tilde{e}_n \rangle e_n$, where $(e_n, \tilde{e}_n)_{n=1,\cdots,N}$ are a biorthogonal basis system for V and $N := \dim(V)$. In this case, using the fact that $E(\tilde{f}) = Pf$ we can estimate the error as follows:

$$\begin{aligned} E(\|\tilde{f} - f\|^2_{L^2}) &= E(\|Pf - f\|^2) + E(\|P(g - f)\|^2) \\ &\le E(\|Pf - f\|^2) + CN\varepsilon^2. \end{aligned}$$

If P is an orthonormal projection, we can assume that $e_n = \tilde{e}_n$ is an orthonormal basis so that $E(\|P(g-f)\|^2) = \sum_n E(|\langle f - g, e_n \rangle|^2) = \sum_n \varepsilon^2$, and therefore the above constant C is equal to 1. Otherwise this constant depends on the "angle" of the projection P. In the above estimation, the first term $E(\|Pf-f\|^2)$ is the *bias* of the estimator. It reflects the approximation property of the space V for the model, and typically decreases with the dimension of V. Note that in the case of a deterministic class Y, it is simply given by $\|Pf-f\|^2$. The second term $CN\varepsilon^2$ represents the *variance* of the estimator which increases with the dimension of V. A good estimator should find an optimal balance between these two terms.

Consider for instance the projection on the multiresolution space V_j, i.e., $\tilde{f} := \sum_{|\lambda|\leq j}\langle g,\tilde{\psi}_\lambda\rangle\psi_\lambda$, together with a deterministic model: the functions f satisfy

$$\|f\|_{H^s} \leq C, \tag{62}$$

where H^s is the Sobolev space of smoothness s. Then we can estimate the bias by the linear approximation estimate in $C2^{-2sj}$ and the variance by $C2^j\varepsilon^2$ since the dimension of V_j adapted to $[0,1]$ is of order 2^j. Assuming an a-priori knowledge on the level ε of the noise, we find that the scale level balancing the bias and variance term is $j(\varepsilon)$ such that $2^{j(\varepsilon)(1+2s)} \sim \varepsilon^{-2}$. We thus select as our estimator

$$\tilde{f} := P_{j(\varepsilon)}g. \tag{63}$$

With such a choice, the resulting estimation error is then bounded by

$$E(\|\tilde{f}-f\|_{L^2}^2) \leq C\varepsilon^{\frac{4s}{1+2s}}. \tag{64}$$

Let us make a few comments on this simple result:

- The convergence rate $4s/(1+2s)$ of the estimator, as the noise level tends to zero, improves with the smoothness of the model. It can be shown that this is actually the optimal or *minimax* rate, in the sense that for any estimation procedure, there always exist an f in the class (62) for which we have $E(\|\tilde{f}-f\|_{L^2}^2) \geq c\varepsilon^{\frac{4s}{1+2s}}$.
- One of the main limitation of the above estimator is that it depends not only on the noise level (which in practice can often be evaluated), but also on the modelizing class itself since $j(\varepsilon)$ depends of s. A better estimator should give an optimal rate for a large variety of function classes.
- The projection $P_{j(\varepsilon)}$ is essentially equivalent to low pass filtering which eliminates the frequencies larger than $2^{j(\varepsilon)}$. The drawbacks of such denoising strategies are well known in practice: while they remove the noise, low-pass filters tend to blur the singularities of the signals, such as the edge in an image. This problem is implicitely reflected in the fact that signals with edges correspond to a value of s which cannot exceed $1/2$ and therefore the convergence rate is at most $\mathcal{O}(\varepsilon)$.

Let us now turn to nonlinear estimation methods based on wavelet thresholding. The simplest thresholding estimator is defined by

$$\tilde{f} := \sum_{|\langle g,\tilde{\psi}_\lambda\rangle|\geq\eta} \langle g,\tilde{\psi}_\lambda\rangle\psi_\lambda, \tag{65}$$

i.e., discarding the coefficients of the data of size less than some $\eta > 0$. Let us remark that the wavelet coefficients of the observed data can be expressed as

$$\langle g,\tilde{\psi}_\lambda\rangle = \langle f,\tilde{\psi}_\lambda\rangle + \varepsilon b_\lambda \tag{66}$$

where the b_λ are gaussian variables. These variables have variance 1 if we assume (which is always possible up to a renormalization) that $\|\tilde{\psi}_\lambda\|_{L^2}^2 = 1$. In the case of an orthonormal bases the b_λ are independent. Therefore the observed coefficients appear as those of the real signal perturbated by an additive noise of level ε. It thus seems at the first sight that a natural choice for a threshold is to simply fix $\eta := \varepsilon$: we can hope to remove most of the noise, while preserving the most significant coefficients of the signal, which is particularly appropriate if the wavelet decomposition of f is sparse.

In order to understand the rate that we could expect from such a procedure, we shall again consider the class of signals described by (56). For a moment, let us assume that we dispose of an *oracle* which gives us the knowledge of those λ such that the wavelet coefficients of the real signal are larger than ε, so that we could build the modified estimator

$$\overline{f} := \sum_{|\langle f,\tilde{\psi}_\lambda\rangle|\geq\varepsilon} \langle g,\tilde{\psi}_\lambda\rangle\psi_\lambda. \tag{67}$$

In this case, $\overline{f}$ can be viewed as the projection Pg of g onto the space $V(f,\varepsilon)$ spanned by the ψ_λ such that $|\langle f,\tilde{\psi}_\lambda\rangle| \geq \varepsilon$, so that we can estimate the error by a sum of bias and variance terms according to

$$\begin{aligned} E(\|\bar{f} - f\|_{L^2}^2) &= \|f - Pf\|^2 + E(\|P(f-g)\|^2) \\ &\leq C[\textstyle\sum_{|\langle f,\tilde{\psi}_\lambda\rangle|\geq\varepsilon} |\langle f,\tilde{\psi}_\lambda\rangle|^2 + \varepsilon^2\#\{\lambda \text{ s.t. } |\langle f,\tilde{\psi}_\lambda\rangle| \geq \varepsilon\}]. \end{aligned}$$

For the bias term, we recognize the nonlinear approximation error which is bounded by $C\varepsilon^{2-\tau}$ according to (58). From the definition of the class (56) we find that the variance term is also bounded by $C\varepsilon^{2-\tau}$. In turn, we obtain for the oracle estimator the convergence rate $\varepsilon^{2-\tau}$. In particular, if we consider the model

$$\|f\|_{B^s_{\tau,\tau}} \leq C, \tag{68}$$

with $1/\tau = 1/2 + s$, we obtain that

$$E(\|\tilde{f} - f\|_{L^2}^2) \leq C\varepsilon^{2-\tau} = C\varepsilon^{\frac{4s}{1+2s}}. \tag{69}$$

Let us again make a few comments:

- In a similar way to approximation rates, nonlinear methods achieve the same estimation rate as linear methods but for much weaker models: the exponent $4s/(1+2s)$ was achieved by the linear estimator for the class (62) which is more restrictive than (56).

- In contrast with the linear estimator, we see that the nonlinear estimator does not need to be tuned according to the value of τ or s. In this sense, it is very robust.
- Unfortunately, (67) is unrealistic since it is based on the "oracle assumption". In practice, we are thresholding according to the values of the observed coefficients $\langle g, \tilde{\psi}_\lambda \rangle = \langle f, \tilde{\psi}_\lambda \rangle + \varepsilon^2 b_\lambda$, and we need to face the possible event that the additive noise $\varepsilon^2 b_\lambda$ severely modifies the position of the observed coefficients with respect to the threshold. Another unrealistic aspect, also in (65), is that one cannot evaluate the full set of coefficients $(\langle g, \tilde{\psi}_\lambda \rangle)_{\lambda \in \nabla}$ which is infinite.

The strategy proposed in [16] solves the above difficulties as follows: a realistic estimator is built by (i) a systematic truncation the estimator (65) above a scale $j(\varepsilon)$ such that $2^{-2\alpha j(\varepsilon)} \sim \varepsilon^2$ for some fixed $\alpha > 0$, and (ii) a choice of threshold slightly above the noise level according to

$$\eta(\varepsilon) := C(\alpha)\varepsilon |\log(\varepsilon)|^{1/2}. \tag{70}$$

It is then possible to prove that the resulting more realistic estimator

$$\tilde{f} := \sum_{|\lambda| \leq j(\varepsilon), |\langle g, \tilde{\psi}_\lambda \rangle| \geq \eta(\varepsilon)} \langle g, \tilde{\psi}_\lambda \rangle \psi_\lambda, \tag{71}$$

has the rate $[\varepsilon |\log(\varepsilon)|^{1/2}]^{\frac{4s}{1+2s}}$ (i.e., almost the same asymptotic performance as the oracle estimator) for the functions which are in both the class (56) and in the Sobolev class H^α. The "minimal" Sobolev smoothness α - which is needed to allow the truncation of the estimator - can be taken arbitrarily close to zero up to a change of the constants in the threshold and in the convergence estimate.

8 Adaptive Numerical Simulation

Numerical simulation is nowadays an essential tool for the understanding of physical processes modelized by partial differential or integral equations. In many instances, the solution of these equations exhibits singularities, resulting in a slower convergence of the numerical schemes as the discretization tends to zero. Moreover, such singularities might be physically significant such as shocks in fluid dynamics or local accumulation of stress in elasticity, and therefore they should be well approximated by the numerical method. In order to maintain the memory size and computational cost at a reasonable level, it is then necessary to use adaptive discretizations which should typically be more refined near the singularities.

In the finite element context, such discretizations are produced by *mesh refinement*: starting from an initial coarse triangulation, we allow further subdivision of certain elements into finer triangles, and we define the discretization space according to this locally refined triangulation. This is of course subject to certain rules, in particular preserving the conformity of the discretization when continuity is required in the finite element space. The use of wavelet bases as an alternative to finite elements is still at its infancy (some first surveys are [6] and [12]), and was strongly motivated by the possibility to produce simple adaptive approximations. In the wavelet context, a more adapted terminology is *space refinement*: we directly produce an approximation space

$$V_\Lambda := \mathrm{Span}\{\psi_\lambda\ ;\ \lambda \in \Lambda\}, \tag{72}$$

by selecting an set Λ which is well adapted to describe the solution of our problem. If N denotes the cardinality of the adapted finite element or wavelet space, i.e., the number of degrees of freedom which are used in the computations, we see that in both cases the numerical solution u_N can be viewed as an adaptive approximation of the solution u in a nonlinear space Σ_N.

A specific difficulty of adaptive numerical simulation is that the solution u is unknown at the start, except for some rough *a-priori* information such as global smoothness. In particular the location and structure of the singularities are often unknown, and therefore the design of an optimal discretization for a prescribed number of degrees of freedom is a much more difficult task than simple compression of fully available data. This difficulty has motivated the development of *adaptive strategies* based on *a-posteriori analysis*, i.e., using the currently computed numerical solution to update the discretization and derive a better adapted numerical solution. In the finite element setting, such an analysis was developed since the 1970's (see [1] or [22]) in terms of *local error indicators* which aim to measure the contribution of each element to the error. The rule of thumb is then to refine the triangles which exhibit the largest error indicators. More recently, similar error indicators and refinement strategies were also proposed in the wavelet context (see [2] and [13]).

Nonlinear approximation can be viewed as a *benchmark* for adaptive strategies: if the solution u can be adaptively approximated in Σ_N with a certain error $\sigma_N(u)$ in a certain norm X, we would ideally like that the adaptive strategy produces an approximation $u_N \in \Sigma_N$ such that the error $\|u - u_N\|_X$ is of the same order as $\sigma_N(u)$. In the case of wavelets, this means that the error produced by the adaptive scheme should be of the same order as the error produced by keeping the N largest coefficients of the exact solution. In most instances unfortunately, such a program cannot be achieved by an adaptive strategy and a more reasonable goal is to obtain an optimal asymptotic rate: if $\sigma_N(u) \leq CN^{-s}$ for some $s > 0$, an *optimal adaptive strategy* should produce an error $\|u - u_N\|_X \leq \tilde{C}N^{-s}$. An additional important aspect is the computational cost to derive u_N: a *computationally optimal strategy* should produce u_N in a number of operation which is proportional to N. A typical instance of computationally optimal algorithm - for a fixed discretization - is the multigrid method for linear elliptic PDE's. It should be noted that very often, the norm X in which one can hope for an optimal error estimate is dictated by the problem at hand: for example, in the case of an elliptic problem, this will typically be a Sobolev norm equivalent to the energy norm (e.g., the H^1 norm when solving the Laplace equation).

Most existing wavelet adaptive schemes have in common the following general structure. At some step n of the computation, a set Λ_n is used to represent the numerical solution $u_{\Lambda_n} = \sum_{\lambda\in\Lambda_n} d_\lambda^n \psi_\lambda$. In the context of an *initial value problem* of the type

$$\partial_t u = E(u), \quad u(x,0) = u_0(x), \tag{73}$$

the numerical solution at step n is typically an approximation to u at time $n\Delta t$ where Δt is the time step of the resolution scheme. In the context of a *stationary problem* of the type

$$F(u) = 0, \tag{74}$$

the numerical solution at step n is typically an approximation to u which should converge to the exact solution as n tends to $+\infty$. In both cases, the derivation of $(\Lambda_{n+1}, u_{\Lambda_{n+1}})$ from $(\Lambda_n, u_{\Lambda_n})$ goes typically in three basic steps:

- *Refinement:* a larger set $\tilde{\Lambda}_{n+1}$ with $\Lambda_n \subset \tilde{\Lambda}_{n+1}$ is derived from an *a-posteriori* analysis of the computed coefficients d_λ^n, $\lambda \in \Lambda_n$.
- *Computation:* an intermediate numerical solution $\tilde{u}_{n+1} = \sum_{\lambda\in\tilde{\Lambda}_{n+1}} d_\lambda^{n+1}\psi_\lambda$ is computed from u_n and the data of the problem.
- *Coarsening:* the smallest coefficients of $\tilde{u}_{n+1}$ are thresholded, resulting in the new approximation $u_{n+1} = \sum_{\lambda\in\Lambda_{n+1}} d_\lambda^{n+1}\psi_\lambda$ supported on the smaller set $\Lambda_{n+1} \subset \tilde{\Lambda}_{n+1}$.

Of course the precise description and tuning of these operations strongly depends on the type of equation at hand, as well as on the type of wavelets which are being used. In the case of linear elliptic problems, it was recently proved in [7] that an appropriate tuning of these three steps results in an optimal adaptive wavelet strategy both in terms of approximation properties and computational time. These results have been extended to more general problems such as saddle points [8] and nonlinear [9]. In the elliptic case, similar results have also been proved in the finite element context : in [3] it is shown that optimal appoximation rates can be achieved by an adaptive mesh refinement algorithm which incorporates coarsening steps that play an analogous role to wavelet thresholding.

9 The Curse of Dimensionality

The three applications that were discussed in the previous sections exploit the sparsity properties of wavelet decompositions for certain classes of functions, or equivalently the convergence properties of nonlinear wavelet approximations of these functions. Nonlinear adaptive methods in such applications are typically relevant if these functions have isolated singularities in which case there might be a substantial gain of convergence rate when switching from linear to nonlinear wavelet approximation. However, a closer look at some simple examples show that this gain tends to decrease for multivariate functions. Consider the L^2-approximation of the characteristic function $f = \chi_\Omega$ of a smooth domain $\Omega \subset [0,1]^d$. Due to the singularity on the boundary $\partial\Omega$, one can easily check that the linear approximation cannot behave better than

$$\sigma_N(f) = \|f - P_j f\|_{L^2} \sim \mathcal{O}(2^{-j/2}) \sim \mathcal{O}(N^{-1/2d}), \tag{75}$$

where $N = \dim(V_j) \sim 2^{dj}$. Turning to nonlinear approximation, we notice that since $\int \tilde{\psi}_\lambda = 0$, all the coefficients d_λ are zero except those such that the support of $\tilde{\psi}_\lambda$ overlaps the boundary. At scale level j there is thus at most $K2^{d-1}j$ non-zero coefficients, where K depends on the support of the ψ_λ and on the $d-1$ dimensional measure of $\partial\Omega$. For such coefficients, we have the estimate

$$|d_\lambda| \le \|\tilde{\psi}_\lambda\|_{L^1} \le C2^{-dj/2}. \tag{76}$$

In the univariate case, i.e., when Ω is a simple interval, the number of non-zero coefficients up to scale j is bounded by jK. Therefore, using N non-zero coefficients at the coarsest levels gives an error estimate with exponential decay

$$\sigma_N(f) \le [\sum_{j \ge N/K} K|C2^{-dj/2}|^2]^{1/2} \le \tilde{C}2^{-dN/2K}, \tag{77}$$

which is a spectacular improvement on the linear rate. In the multivariate case, the number of non-zero coefficients up to scale j is bounded by $\sum_{l=0}^{j} K2^{(d-1)l}$ and thus by $\tilde{K}2^{(d-1)j}$. Therefore, using N non-zero coefficients at the coarsest levels gives an error estimate

$$\sigma_N(f) \le [\sum_{\tilde{K}2^{(d-1)j} \ge N} K2^{(d-1)j}|C2^{-dj/2}|^2]^{1/2} \le \tilde{C}N^{-1/(2d-2)}, \tag{78}$$

which is much less of an improvement. For example, in the 2D case, we only go from $N^{-1/4}$ to $N^{-1/2}$ by switching to nonlinear wavelet approximation.

This simple example illustrates the *curse of dimensionality* in the context of nonlinear wavelet approximation. The main reason for the degradation of the approximation rate is the large number $K2^{(d-1)j}$ of wavelets which are needed to refine the boundary from level j to level $j+1$. On the other hand, if we view the boundary itself as the graph of a smooth function, it is clear that approximating this graph with accuracy 2^{-j} should require much less parameters than $K2^{(d-1)j}$. This reveals the fundamental limitation of wavelet bases: they fail to exploit the smoothness of the boundary and therefore cannot capture the simplicity of f in a small number of parameters. Another way of describing this limitation is by remarking that nonlinear wavelet approximation allows local refinement of the approximation, but imposes some *isotropy* in this refinement process. In order to capture the boundary with a small number of parameters, one would typically need to refine more in the normal direction than in the tangential directions, i.e., apply *anisotropic local refinement.*

In this context, other approximation tools outperform wavelet bases: it is easy to check that the use of piecewise constant functions on an adaptive partition of N triangles in $2D$ will produce the rate $\sigma_N(f) \sim \mathcal{O}(N^{-1})$, precisely because one is allowed to use arbitrarily anisotropic triangles to match the boundary. In the case of rational functions it is conjectured an even more spectacular result: if $\partial\Omega$ is C^∞, then $\sigma_N(f) \le C_r N^{-r}$ for any $r > 0$. These remarks reveal that, in contrast to the 1D case, free triangulations or rational approximation outperform N-term approximation, and could be thought as a better tool in view of applications such as those which were discussed throughout this paper. This is not really true in practice: in numerical simulation, rational functions are difficult to use and free triangulations are often limited by shape constraints which restricts their anisotropy, and both methods are not being used in statistical estimation or data compression, principally due to the absence of fast and robust algorithms which would produce an optimal adaptive approximation in a similar manner as wavelet thresholding. The development of new approximation and representation tools, which could both capture anisotropic features such as edges with a very small number of parameters and be implemented by fast and robust procedures, is currently the object of active research. A significant breakthrough was recently achieved by Donoho and Candes who developed representations into *ridglet* bases which possess the scale-space localization of wavelets together with some directional selection. Such bases allow for example to recover with a simple thresholding procedure the rate $\mathcal{O}(N^{-1})$ for a bivariate function which is smooth except along a smooth curve of discontinuity.

References

1. Babushka, I. and W. Reinhbolt (1978) *A-posteriori analysis for adaptive finite element computations*, SIAM J. Numer. Anal. 15, 736–754.
2. Bertoluzza, S. (1995) *A posteriori error estimates for wavelet Galerkin methods*, Appl. Math. Lett. 8, 1–6.
3. Binev P., W. Dahmen and R. DeVore (2002), *Adaptive finite element methods with convergence rate*, preprint IGPM-RWTH Aachen, to appear in Numer. Math..
4. Candes, E. and D. Donoho (1999), *Ridgelets: a key to higher-dimensional intermittency ?*, Phil. Trans. Roy. Soc..
5. Ciarlet, P.G. (1991), *Basic error estimates for the finite element method*, Handbook of Numerical Analysis, vol II, P. Ciarlet et J.-L. Lions eds., Elsevier, Amsterdam.
6. Cohen, A. (2001) *Wavelets in numerical analysis*, Handbook of Numerical Analysis, vol. VII, P.G. Ciarlet and J.L. Lions, eds., Elsevier, Amsterdam.

7. Cohen, A., W. Dahmen and R. DeVore (2000) *Adaptive wavelet methods for elliptic operator equations - convergence rate*, Math. Comp. **70**, 27–75.
8. Cohen, A., W. Dahmen and R. DeVore (2002) *Adaptive wavelet methods for elliptic equations - beyond the elliptic case*, Found. of Comp. Math. **2**, 203–245.
9. Cohen, A., W. Dahmen and R. DeVore (2003) *Adaptive wavelet schemes for nonlinear variational problems*, preprint IGPM-RWTH Aachen, to appear in SIAM J. Numer. Anal..
10. Cohen, A., W. Dahmen, I. Daubechies and R. DeVore (2001) *Tree approximation and optimal encoding*, Appl. Comp. Harm. Anal. **11**, 192–226.
11. Cohen, A., R. DeVore, P. Petrushev and H. Xu (1999), *Nonlinear approximation and the space* $BV(\mathbb{R}^2)$, Amer. Jour. of Math. 121, 587-628.
12. Dahmen, W. (1997) *Wavelet and multiscale methods for operator equations*, Acta Numerica 6, 55–228.
13. Dahmen, W., S. Dahlke, R. Hochmuth and R. Schneider (1997) *Stable multiscale bases and local error estimation for elliptic problems*, Appl. Numer. Math. 23, 21–47.
14. Daubechies, I. (1992) *Ten lectures on wavelets*, SIAM, Philadelphia.
15. DeVore, R. (1997) *Nonlinear Approximation*, Acta Numerica 51–150.
16. Donoho, D., I. Johnstone, G. Kerkyacharian and D. Picard (1992) *Wavelet shrinkage: asymptotia ? (with discussion)*, Jour. Roy. Stat. Soc. 57, Serie B, 301-369.
17. DeVore R. and X. M. Yu (1990) *Degree of adaptive approximation*, Math. Comp. 55, 625–635.
18. Mallat, S. (1998) *A wavelet tour of signal processing*, Academic Press, New York.
19. Petrushev, P. (1988) *Direct and converse theorems for spline and rational approximation and Besov spaces*, in *Function spaces and applications*, M. Cwikel, J. Peetre, Y. Sagher and H. Wallin, eds., Lecture Notes in Math. 1302, Springer Verlag, Berlin, 363–377.
20. Shapiro, J. (1993) *Embedded image coding using zerotrees of wavelet coefficients*, IEEE Signal Processing 41, 3445–3462.
21. Temlyakov, V. (1998) *Best N-term approximation and greedy algorithms*, Adv. Comp. Math. **8**, 249–265.
22. Verfürth, R. (1994) *A-posteriori error estimation and adaptive mesh refinement techniques*, Jour. Comp. Appl. Math. 50, 67–83.

Multiscale and Wavelet Methods for Operator Equations

Wolfgang Dahmen

Institut für Geometrie und Praktische Mathematik, RWTH Aachen, Germany
dahmen@igpm.rwth-aachen.de

Summary. These notes are to bring out some basic mechanisms governing wavelet methods for the numerical treatment of differential and integral equations. Some introductory examples illustrate the quasi–sparsity of wavelet representations of functions and operators. This leads us to identify the key features of general wavelet bases in the present context, namely locality, cancellation properties and norm equivalences. Some analysis and construction principles regarding these properties are discussed next. The scope of problems to which these concepts apply is outlined along with a brief discsussion of the principal obstructions to an efficient numerical treatment. This covers elliptic boundary value problems as well as saddle point problems. The remainder of these notes is concerned with a new paradigm for the adaptive solution of such problems. It is based on an equivalent formulation of the original variational problem in wavelet coordinates. Combining the well-posedness of the original problem with the norm equivalences induced by the wavelet basis, the transformed problem can be arranged to be well–posed in the Euclidean metric. This in turn allows one to devise convergent iterative schemes for the infinite dimensional problem over ℓ_2. The numerical realization consists then of the adaptive application of the wavelet representations of the involved operators. Such application schemes are described and the basic concepts for analyzing their computational complexity, rooted in nonlinear approximation, are outlined. We conclude with an outlook on possible extensions, in particular, to nonlinear problems.

1 Introduction

These lecture notes are concerned with some recent developments of wavelet methods for the numerical treatment of certain types of operator equations. A central theme is a new approach to discretizing such problems that differs from conventional concepts in the following way. Wavelets are used to transform the problem into an equivalent one which is well-posed in ℓ_2. The solution of the latter one is based on concepts from nonlinear approximation. This part of the material is taken from recent joint work with A. Cohen, R. DeVore and also with S. Dahlke and K. Urban.

The notes are organized as follows. Some simple examples are to identify first some key features of multiscale bases that will motivate and guide the subsequent discussions. Some construction and analysis principles are then reviewed that can be used to construct bases with the relevant properties for realistic application settings. The scope of these applications will be detailed next. After these preparations the remaining part is devoted to the discussion of adaptive solution techniques.

2 Examples, Motivation

Rather than *approximating* a given function by elements of a suitable finite dimensional space the use of *bases* aims at *representing* functions as expansions thereby identifying the function with an array of coefficients – its *digits*. The following considerations indicate why wavelet bases offer particularly favorable digit representations.

2.1 Sparse Representations of Functions, an Example

Consider the space $L_2([0,1])$ of all square integrable functions on the interval $[0,1]$. For a given dyadic mesh of mesh size 2^{-j} piecewise constant approximations on such meshes are conveniently formed by employing dilates and translates of the indicator function of $[0,1]$.

$$\phi(x) := \chi_{[0,1)}(x)$$

$$\phi_{j,k} := 2^{j/2}\phi\left(2^j \cdot -k\right), \quad k = 0, \dots 2^j - 1,$$

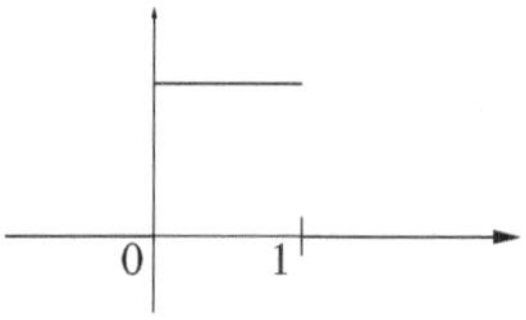

In fact, denoting by $\langle f, g\rangle = \langle f, g\rangle_{[0,1]} := \int\limits_0^1 f(x)g(x)dx$ the standard inner product on $[0,1]$,

$$P_j(f) := \sum_{k=0}^{2^j-1} \langle f, \phi_{j,k}\rangle \phi_{j,k}$$

is the orthogonal projection of f onto the space $S_j := \operatorname{span}\{\phi_{j,k} : k = 0, \dots, 2^j - 1\}$. Figure 1 displays such approximations for several levels of resolution.

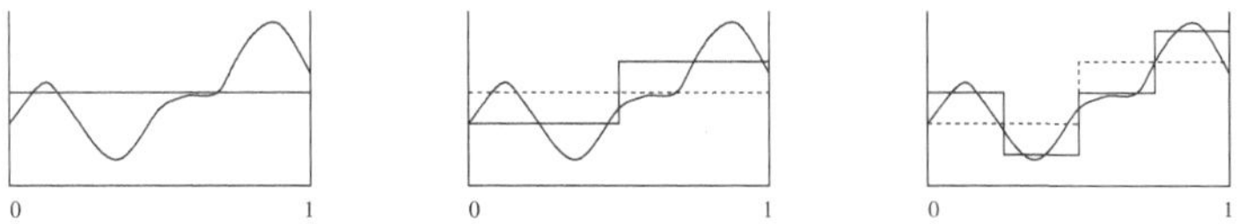

Fig. 1. Piecewise constant approximations to f

If a chosen level of resolution turns out to be insufficient, a naive approach would be to recompute for a smaller mesh size. More cleverly one can avoid wasting prior efforts by monitoring the *difference* between successive levels of resolution as indicated by Figure 2

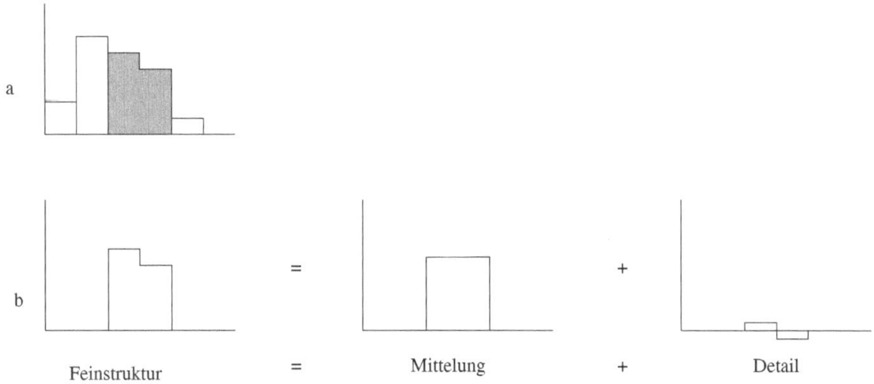

Fig. 2. Splitting averages and details

To encode the split off details, in addition to the averaging profile ϕ a second *oscillatory profile* is needed. Defining

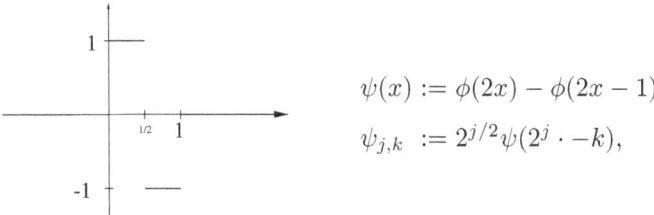

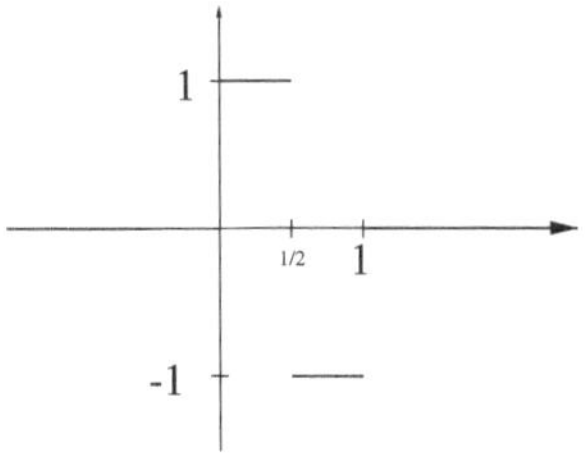

$$\psi(x) := \phi(2x) - \phi(2x-1)$$
$$\psi_{j,k} := 2^{j/2}\psi(2^j \cdot -k),$$

it is not hard to show that

$$(P_{j+1} - P_j)f = \sum_{k=0}^{2^j-1} d_{j,k}(f)\psi_{j,k} \quad \text{where} \quad d_{j,k}(f) = \langle f, \psi_{j,k}\rangle. \tag{1}$$

Thus, due to the denseness of piecewise constants in $L_2([0,1])$, the telescoping expansion converges strongly to f, so that

$$f = P_0 f + \sum_{j=1}^{\infty}(P_j - P_{j-1})f = \sum_{j=-1}^{\infty}\sum_{k=0}^{2^j-1} d_{j,k}(f)\psi_{j,k} =: \boldsymbol{d}(f)^T\Psi \qquad (2)$$

is a representation of f.
Norm equivalence: Due to the fact that

$$\Psi = \{\phi_{0,0}\} \cup \{\psi_{j,k} : k = 0, \ldots, 2^j - 1\}$$

is an *orthonormal* collection of functions – basis in this case – one has

$$\|f\|_{L_2} = \left(\sum_{j=0}^{\infty} \|(P_j - P_{j-1})f\|_{L_2}^2\right)^{1/2} = \|\boldsymbol{d}(f)\|_{\ell_2}. \qquad (3)$$

Since therefore small changes in $\boldsymbol{d}(f)$ cause only equally small changes in f and vice versa, we have a particularly favorable digit representation of f. Ψ is called the *Haar-basis* – a first simple example of a *wavelet basis.*
Vanishing moment: While summability in (3) implies that the detail or *wavelet coefficients* $d_{j,k}(f)$ eventually tend to zero – just as in Fourier expansions, it is interesting to see to what extent – in contrast to Fourier expansions – their size conveys *local* information about f. Since the $\psi_{j,k}$ are orthogonal to constants, one readily concludes

$$|d_{j,k}(f)| = \inf_{c\in\mathbb{R}} |\langle f - c, \psi_{j,k}\rangle| \le \inf_{c\in\mathbb{R}} \|f - c\|_{L_2(I_{j,k})} \le 2^{-j}\|f'\|_{L_2(I_{j,k})}. \qquad (4)$$

Thus $d_{j,k}(f)$ is mall when f is smooth on the support $I_{j,k}$ of $\psi_{j,k}$.

If all wavelet coefficients $d_{j,k}(f)$ were known, keeping the N largest among them and replacing all others by zero, would provide an approximation to f involving only N terms that minimizes, in view of (3), the L_2-error among all other competing N term approximations from the Haar system. These N terms would give rise to a possibly highly nonuniform mesh for the corresponding piecewise constant approximation. Moreover, (4) tells us that in regions where f has a small first derivative the mesh size would be relatively coarse, while small intervals are encountered where f varies strongly. This ability to provide *sparse approximate representations of functions* is a key feature of wavelet bases.

Wavelet transform: Computing wavelet coefficients directly through quadrature poses difficulties. Since the support of low level wavelets is comparable to the domain, a sufficiently accurate quadrature would be quite expensive. A remedy is offered by the following strategy which can be used when the wavelet coefficients of interest have at most some highest level J say. The accurate computation of the *scaling function coefficients* $c_{J,k} := \langle f, \phi_{J,k}\rangle$, $k = 0, \ldots, 2^J - 1$, is much less expensive, due to their uniformly small support. The transformation from the array $\mathbf{c}_J$ into the array of wavelet coefficients

$\boldsymbol{d}^J = (\mathbf{c}_0, \boldsymbol{d}_0, \ldots, \boldsymbol{d}_{J-1})$, $\boldsymbol{d}_j := \{d_{j,k}(f) : k = 0, \ldots, 2^j - 1\}$ is already implicit in (1). Its concrete form can be derived from the *two-scale relations:*

$$\phi_{j,k} = \frac{1}{\sqrt{2}}(\phi_{j+1,2k} + \phi_{j+1,2k+1}), \quad \psi_{j,k} := \frac{1}{\sqrt{2}}(\phi_{j+1,2k} - \phi_{j+1,2k+1}), \tag{5}$$

$$\phi_{j+1,2k} = \frac{1}{\sqrt{2}}(\phi_{j,k} + \psi_{j,k}), \quad \phi_{j+1,2k+1} = \frac{1}{\sqrt{2}}(\phi_{j,k} - \psi_{j,k}). \tag{6}$$

In fact, viewing the array of basis functions as a vector (in some fixed unspecified order) and adopting in the following the shorthand notation

$$\sum_{k=0}^{2^{j+1}-1} c_{j+1,k}\phi_{j+1,k} =: \Phi_{j+1}^T \mathbf{c}_{j+1},$$

Figure 2 suggests to write the fine scale scaling function representation as a coarse scale representation plus detail: $\Phi_{j+1}^T \mathbf{c}_{j+1} = \Phi_j^T \mathbf{c}_j + \Psi^T \boldsymbol{d}_j$. One easily derives from (5) that the coefficients are interrelated as follows

$$c_{j,k} = \frac{1}{\sqrt{2}}(c_{j+1,2k} + c_{j+1,2k+1}), \quad d_{j,k} = \frac{1}{\sqrt{2}}(c_{j+1,2k} - c_{j+1,2k+1}),$$

$$c_{j+1,2k} = \frac{1}{\sqrt{2}}(c_{j,k} + d_{j,k}), \quad c_{j+1,2k+1} = \frac{1}{\sqrt{2}}(c_{j,k} - d_{j,k}).$$

This leads to the following cascadic transforms whose structure is shared also by more complex wavelet bases.

Fast (Orthogonal) Transform:
$\mathbf{T}_J : \boldsymbol{d}^J := (\mathbf{c}_0, \boldsymbol{d}_0, \ldots, \boldsymbol{d}_{J-1}) \to \mathbf{c}_J$:

$$\begin{array}{ccccccccccc} \mathbf{c}_0 & \to & \mathbf{c}_1 & \to & \mathbf{c}_2 & \to \cdots \to & \mathbf{c}_{J-1} & \to & \mathbf{c}_J \\ & \nearrow & & \nearrow & & & \nearrow & & \nearrow \\ \boldsymbol{d}_0 & & \boldsymbol{d}_1 & & \boldsymbol{d}_2 & \cdots & & \boldsymbol{d}_{J-1} & \end{array}$$

Inverse transform: $\mathbf{T}_J^{-1} = \mathbf{T}_J^T : \mathbf{c}_J \to \boldsymbol{d}^J$:

$$\begin{array}{ccccccccc} \mathbf{c}_J & \to & \mathbf{c}_{J-1} & \to & \mathbf{c}_{J-2} & \to \cdots \to & \mathbf{c}_1 & \to & \mathbf{c}_0 \\ & \searrow & & \searrow & & & \searrow & & \searrow \\ & & \boldsymbol{d}_{J-1} & & \boldsymbol{d}_{J-2} & \cdots & & \boldsymbol{d}_1 & & \boldsymbol{d}_0 \end{array}.$$

Obviously, both transforms $\mathbf{T}_J$ and $\mathbf{T}_J^{-1}$ are efficient in the sense that the computational effort stays proportional to the size of $\mathbf{c}_J$. However, one should already note at this point that this strategy does not fit with the case when f admits actually a much sparser approximation in wavelet coordinates. Finding alternative strategies will be an important subject of subsequent discussions.

2.2 (Quasi-) Sparse Representation of Operators

Wavelet bases not only offer nearly sparse representations of functions but also of operators from a wide class. Let us explain first what is meant by *standard wavelet representation* of a linear operator for the example of the Haar basis. Suppose for simplicity that $\mathcal{L}$ takes $L_2([0,1])$ into itself. Expanding f in the wavelet basis, applying $\mathcal{L}$ to each basis function and expanding the resulting images again in the wavelet basis, yields

$$\mathcal{L}f = \sum_{j,k} \langle f, \psi_{j,k}\rangle \mathcal{L}\psi_{j,k} = \sum_{j,k}\left(\sum_{l,m}\langle \mathcal{L}\psi_{j,k}, \psi_{l,m}\rangle \psi_{l,m}\right)\langle f, \psi_{j,k}\rangle = \Psi^T \mathbf{L}\boldsymbol{d},$$

where

$$\mathbf{L} := \left(\langle \psi_{j,k}, \mathcal{L}\psi_{l,m}\rangle\right)_{(j,k),(l,m)} =: \langle \Psi, \mathcal{L}\Psi\rangle, \quad \boldsymbol{d} := \left(\langle \psi_{j,k}, f\rangle\right)_{(j,k)} =: \langle \Psi, f\rangle. \tag{7}$$

When $\mathcal{L}$ is *local* like a differential operator many of the entries vanish. However, when $\mathcal{L}$ is *global* like an integral operator, in principle, all the entries of $\mathbf{L}$ could be nonzero so that finite sections would be densely populated matrices with corresponding adverse effects on computational complexity. It was observed in [13] that for certain singular integral operators $\mathbf{L}$ is almost sparse in the sense that many entries became very small. Let us quantify this statement for the following example of the *the Hilbert Transform*

$$(\mathcal{L}f)(x) := \frac{1}{\pi} p.v. \int_{\mathbb{R}} \frac{f(y)}{x-y}\, dy.$$

Using Taylor expansion of the kernel and again the fact that the wavelets are orthogonal to constants, one obtains

$$\begin{aligned}
\pi \mathbf{L}_{(j,k),(l,m)} &= \int_{2^{-j}k}^{2^{-j}(k+1)} \left\{ \int_{2^{-l}m}^{2^{-l}(m+1)} \left(\frac{1}{x-y} - \frac{1}{x-2^{-l}m}\right)\psi_{l,m}(y)\, dy \right\} \psi_{j,k}(x)\, dx \\
&= \int_{2^{-l}m}^{2^{-l}(m+1)} \left\{ \int_{2^{-j}k}^{2^{-j}(k+1)} \frac{(y-2^{-l}m)}{(x-y_{l,m})^2}\psi_{j,k}(x)\, dx \right\} \psi_{l,m}(y)\, dy \\
&= \int_{2^{-l}m}^{2^{-l}(m+1)} \left\{ \int_{2^{-j}k}^{2^{-j}(k+1)} \left(\frac{(y-2^{-l}m)}{(x-y_{l,m})^2} - \frac{(y-2^{-l}m)}{(2^{-j}k-y_{l,m})^2}\right)\psi_{j,k}(x)\, dx \right\} \\
&\quad \times \psi_{l,m}(y)\, dy.
\end{aligned}$$

Since $\int_{\mathbb{R}} |\psi_{j,k}(x)|\,dx \le 2^{-j/2}$ one therefore obtains, for instance when $l \le j$,

$$\pi|\mathbf{L}_{(j,k),(m,l)}| \lesssim 2^{-(l+j)\frac{3}{2}}|2^{-j}k - 2^{-l}m|^{-3} = \frac{2^{-\frac{3}{2}|j-l|}}{|k - 2^{j-l}m|^3},$$

which says that the entries decay exponentially with increasing difference of levels and also polynomially with increasing spatial distance of the involved wavelets. A more general estimate for general wavelet bases will later play an important role.

2.3 Preconditioning

The next example addresses an effect of wavelets which is related to *preconditioning*. As a simple example consider the boundary value problem

$$-u'' - f \quad \text{on } [0,1], \quad u(0) = u(1) = 0,$$

whose weak formulation requires finding $u \in H_0^1([0,1])$ such that

$$\langle u', v' \rangle = \langle f, v \rangle, \quad v \in H_0^1([0,1]).$$

Here $H_0^1([0,1])$ denotes the space of functions in $L_2([0,1])$ whose first derivative is also in $L_2([0,1])$ and which vanish at 0 and 1. Note that this formulation requires less regularity than the original strong form. Therefore we can employ piecewise linear continuous functions as building blocks for representing approximate solutions. Specifically, denoting by S_j this time the span of the functions $\phi_{j,k} = 2^{j/2}\phi\left(2^j \cdot -k\right)$, $k = 1, \ldots, 2^j - 1$, which are dilates and translates of the classical *hat function* $\phi(x) := (1 - |x|)_+$, the corresponding *Galerkin scheme* requires finding $u_j \in S_j$ such that

$$\langle u_j', v' \rangle = \langle f, v \rangle, \quad v \in S_j. \tag{8}$$

Clearly, making the ansatz $u_J = \sum_{k=1}^{2^J-1} u_{J,k}\phi_{J,k}$, gives rise to a linear system of equations

$$\boldsymbol{A}_J\mathbf{u}_J = \mathbf{f}_J, \quad \text{where } \boldsymbol{A}_J := \langle \Phi'_J, \Phi'_J \rangle, \ \ \mathbf{f}_J := \langle \Phi_j, f \rangle. \tag{9}$$

Here and below we denote for any two countable collections $\Theta \subset Y, \Xi \subset X$ and any bilinear form $c(\cdot,\cdot)$ on $Y \times X$ by $c(\Theta, \Xi)$ the *matrix* $(c(\theta,\xi))_{\theta\in\Theta,\xi\in\Xi}$. In particular, as used before, $\langle \Phi_j, f \rangle$ is then a column vector.

Although in practice one would not apply an iterative scheme for the solution of the particular system (9), it serves well to explain what will be relevant for more realistic multidimensional problems. The performance of an iterative scheme for a symmetric positive system is known to depend on the *condition number* of that system which in this case is the quotient of the maximal and minimal eigenvalue. Although it will be explained later in a little bit more detail, what can be said about the condition numbers for problems of the above type (see Section 10.3), it should suffice for the moment to note that the condition numbers grow like h^{-2} (here 2^{2J} for $h = 2^{-J}$) for a given mesh size h, which indeed adversely affects the performance of the iteration. This growth can be made plausible by noting that the second derivative treats highly oscillatory functions very differently from slowly varying functions. Since the trial space S_J contains functions with frequency ranging between one and 2^J, this accounts for the claimed growth. This motivates to split the space S_J into direct summands consisting of functions with fixed frequency. On each such summand the differential operator would be well conditioned. If the summands were well separated with respect to the *energy inner product* $a(v,w) := \langle v', w' \rangle$ this should enable one to *precondition* the whole problem. This is the essence of *multilevel preconditioners* and can be illustrated well in the present simple case.

The key is again that the main building block, the hat function, is *refinable* in the sense that

$$\phi(x) = \tfrac{1}{2}\phi(2x+1) + \phi(2x) + \tfrac{1}{2}\phi(2x-1)$$

$$\phi_{j,k} = \tfrac{1}{\sqrt{2}}\left(\tfrac{1}{2}\phi_{j+1,2k-1} + \phi_{j+1,2k} + \tfrac{1}{2}\phi_{j+1,2k+1}\right),$$

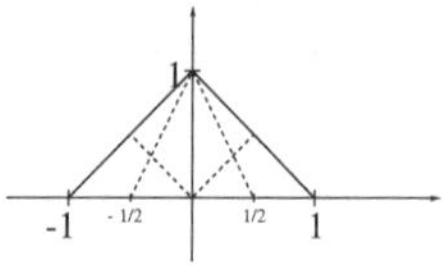

which means that the trial spaces S_j are nested. As in the case of the Haar basis one builds now an alternative *multilevel basis* for S_J by retaining basis functions on lower levels and adding additional functions in a complement space between two successive spaces S_j and S_{j+1}. The simplest complement spaces are those spanned by the hat functions for the *new nodes* on the next higher level, see Figure 3. The resulting multilevel basis has become known as *hierarchical basis* [75].

$$\Psi_j := \{\psi_{j,k} := \phi_{j+1,2k+1} : k = 0, \ldots, 2^j - 1\}$$

$$S_{j+1} = S_j \oplus \mathrm{span}\,(\Psi_j)$$

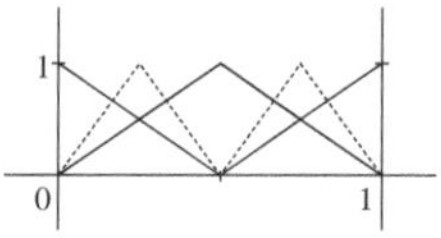

Fig. 3. Hierarchical basis

Denoting the Haar wavelets for the sake of distinction here by $\psi^H_{j,k}$, one can check that $\frac{d}{dx}\psi_{j,k}(x) = \frac{d}{dx}\phi_{j+1,2k+1}(x) = 2^{j+\frac{3}{2}}\psi^H_{j,k}(x)$ so that

$$\langle \frac{d}{dx}\psi_{j,k}, \frac{d}{dx}\psi_{l,m} \rangle = 2^{2j+3}\delta_{(j,k),(l,m)}. \tag{10}$$

Thus the stiffness matrix relative to the hierarchical basis is even *diagonal* and (in a slight abuse of the term) a simple diagonal scaling would yield *uniformly bounded* condition numbers independent of the mesh size.

2.4 Summary

So far we have seen some basic effects of two features, namely *Vanishing Moments – Cancellation Property (CP)* of wavelets which entail:

- sparse representations of functions;
- sparse representations of operators;

as well as *norm equivalences (NE)* which imply:

- tight relations between functions and their digit representation;
- well conditioned systems.

Of course, the above examples are very specific and the question arise to what extend the above properties are needed in less simple situations. Specifically, what is the role of orthogonality? Diagonalization is certainly much stronger than preconditioning.

We will extract next somewhat weaker features of multilevel bases that have a better chance to be practically feasible in a much more general framework and will then point out why these features are actually sufficient for the purposes indicated above.

3 Wavelet Bases – Main Features

The objective is to develop flexible concepts for the construction of multiscale bases on one hand for realistic domain geometries Ω (bounded Euclidean domains, surfaces, manifolds) as well as for interesting classes of operator equations so as to exploit in essence the features exhibited by the above examples.

3.1 The General Format

We postpone discussing the concrete construction of such bases but are content for the moment with describing their general format, always keeping the above examples as guide line in mind. We consider collections $\Psi = \{\psi_\lambda : \lambda \in \mathcal{J}\} \subset L_2(\Omega)$ of functions – wavelets – that are normalized in L_2, i.e. $\|\psi_\lambda\|_{L_2} = 1$, $\lambda \in \mathcal{J}$, where $\dim \Omega = d$. Here $\mathcal{J} = \mathcal{J}_\phi \cup \mathcal{J}_\psi$ is an infinite index set where: $\#\mathcal{J}_\phi < \infty$ representing the "scaling functions", like the box or hat functions above, living on the coarsest scale. For Euclidean domains these functions will span polynomials up to some order which will be called the order of the basis Ψ. The indices in $\mathcal{J}_\psi$ represent the "true" wavelets spanning complements between refinement levels. Each index $\lambda \in \mathcal{J}$

encodes different types of information, namely the *scale* $j = j(\lambda) = |\lambda|$, the *spatial location* $k = k(\lambda)$ and the *type* $e = e(\lambda)$ of the wavelet. Recall that e.g. for tensor product constructions one has $2^d - 1$ different types of wavelets associated with each spatial index k. For $d = 2$ one has, for instance, $\psi_\lambda(x,y) = 2^j\psi^{1,0}(2^j(x,y) - (k,l)) = 2^{j/2}\psi(2^jx - k)2^{j/2}\phi(2^jy - l)$. We will explain later what exactly qualifies Ψ as a *wavelet basis* in our context.

3.2 Notational Conventions

As before it will be convenient to view a collection Ψ as an (infinite) vector (with respect to some fixed but unspecified order of the indices in $\mathcal{J}$). We will always denote by $\mathbf{D} = \operatorname{diag}(w_\lambda : \lambda \in \mathcal{J})$ a diagonal matrix and write $\mathbf{D} = \mathbf{D}^s$ when the diagonal entries are $w_\lambda = 2^{s|\lambda|}$, a frequently occurring case. Thus $\mathbf{D}^{-s}\Psi = \{2^{-s|\lambda|}\psi_\lambda\}$ will often stand for a *scaled* wavelet basis. Arrays of wavelet coefficients will be denoted by boldface letters like $\mathbf{v} = \{v_\lambda\}_{\lambda\in\mathcal{J}}$, $\boldsymbol{d}, \mathbf{u}, \ldots$,. The above shorthand notation allows us then to write wavelet expansions as $\boldsymbol{d}^T\Psi := \sum_{\lambda\in\mathcal{J}} d_\lambda\psi_\lambda$. In the following the notation $\boldsymbol{d}$ for the wavelet coefficients indicates expansions with respect to the *unscaled* L_2-normalized basis while $\mathbf{v}, \mathbf{u}$ will typically be associated with a scaled basis. Recall from (9) that (generalized) Gramian matrices are for any bilinear form $c(\cdot,\cdot) : X \times Y \to \mathbb{R}$ and (countable) collections $\Xi \subset X, \Theta \subset Y$ denoted by

$$c(\Xi,\Theta) := (c(\xi,\theta))_{\xi\in\Xi,\theta\in\Theta}, \quad \langle\Psi, f\rangle = (\langle f, \psi_\lambda\rangle)^T_{\lambda\in\mathcal{J}},$$

where the right expression is viewed as a column vector.

3.3 Main Features

The main features of wavelet bases for our purposes can be summarized as follows:

- Locality **(L)**;
- Cancellation Properties **(CP)**;
- Norm Equivalences **(NE)**;

Some comments are in order.

Locality (L): By this we mean that the wavelets are compactly supported and that the supports scale as follows.

$$\Omega_\lambda := \operatorname{supp}\psi_\lambda, \quad \operatorname{diam}(\Omega_\lambda) \sim 2^{-|\lambda|}. \tag{11}$$

Of course, one could replace 2 by some $\rho > 1$. But this is merely a technical point and will be dismissed for simplicity.

Cancellation Property (CP) of Order $\tilde{m}$: This means that integration against a wavelet annihilates smooth parts which can be expressed in terms of inequalities of the following type:

$$|\langle v, \psi_\lambda \rangle| \lesssim 2^{-|\lambda|(\tilde{m}+\frac{d}{2}-\frac{d}{p})} |v|_{W_p^{\tilde{m}}(\Omega_\lambda)}, \quad \lambda \in \mathcal{J}_\psi, \tag{12}$$

see (4) for $\tilde{m} = 1, d = 1, p = 2$. The most familiar sufficient (in fact equivalent) condition for (12) for wavelets on Euclidean domains is that the wavelets have *vanishing polynomial moments* of order $\tilde{m}$ which means that

$$\langle P, \psi_\lambda \rangle = 0, \quad P \in \mathbb{P}_{\tilde{m}},\ \lambda \in \mathcal{J}_\psi, \tag{13}$$

where $\mathbb{P}_{\tilde{m}}$ denotes the space of all polynomials of (total) degree $\tilde{m} - 1$ (order $\tilde{m}$). In fact, the same argument as in (4) yields for $\frac{1}{p} + \frac{1}{p'} = 1$

$$\begin{aligned} |\langle v, \psi_\lambda \rangle| &= \inf_{P \in \mathbb{P}_{\tilde{m}}} |\langle v - P, \psi_\lambda \rangle| \leq \inf_{P \in \mathbb{P}_{\tilde{m}}} \|v - P\|_{L_p(\Omega_\lambda)} \|\psi_\lambda\|_{L_{p'}}, \\ &\lesssim 2^{-|\lambda|(\frac{d}{2}-\frac{d}{p})} \inf_{P \in \mathbb{P}_{\tilde{m}}} \|v - P\|_{L_p(\Omega_\lambda)}, \end{aligned}$$

where we have used that

$$\|\psi_\lambda\|_{L_{p'}} \sim 2^{|\lambda|\left(\frac{d}{p'}-\frac{d}{2}\right)} \sim 2^{|\lambda|(\frac{d}{2}-\frac{d}{p})}. \tag{14}$$

Now one can invoke standard Whitney-type estimates for *local polynomial approximation* of the form

$$\inf_{P \in \mathbb{P}_k} \|v - P\|_{L_p(\Omega)} \lesssim (\operatorname{diam} \Omega)^k |v|_{W_p^k(\Omega)} \tag{15}$$

to confirm (12) in this case (for more details about (15) see Section 10.2).

Norm Equivalences (NE): This is perhaps the most crucial point. It should be emphasized that, in comparison with conventional numerical schemes, wavelet concepts aim at more than just finite dimensional approximations of a given or searched for function but rather at its *representation* in terms of an array of wavelet coefficients – the *digits* of the underlying function. The tighter the interrelation between function and digits is, the better. We will make heavy use of the following type of such interrelations. For some $\gamma, \tilde{\gamma} > 0$ and $s \in (-\tilde{\gamma}, \gamma)$ there exist positive bounded constants c_s, C_s such that

$$c_s \|\mathbf{v}\|_{\ell_2} \leq \|\mathbf{v}^T \mathbf{D}^{-s} \Psi\|_{H^s} \leq C_s \|\mathbf{v}\|_{\ell_2}, \quad \mathbf{v} \in \ell_2, \tag{16}$$

where for $s \geq 0$ the space H^s will always stand for a closed subspace of the Sobolev space $H^s(\Omega)$, defined e.g. by imposing homogeneous boundary conditions on parts or all of the boundary of Ω, i.e., $H_0^s(\Omega) \subseteq H^s \subseteq H^s(\Omega)$. For $s < 0$, the space H^s is always understood as the dual space $H^s := (H^{-s})'$.

(16) means that the scaled basis $\mathbf{D}^{-s}\Psi$ is a *Riesz-basis* for H^s, i.e., every element in H^s has a unique expansion satisfying (16). Thus small changes in

the coefficients can cause only small changes in the function and vice versa, which is obviously a desirable feature with respect to stability.

We will discuss next some ramifications of (16). The first observation is that (16) entails further norm equivalences for other Hilbert spaces. The following example addresses the issue of *robustness* of such norm equivalences. The point here is that these relations can often be arranged to be independent of some parameter in an *energy inner product* and the corresponding norm, a fact that will be, for instance, relevant in the context of preconditioning.

Remark 1 *Define* $\|v\|^2_{\mathcal{H}_\epsilon} := \epsilon\langle\nabla v, \nabla v\rangle + \langle v, v\rangle$ *and consider the diagonal matrix* $\mathbf{D}_\epsilon := \left((1+\sqrt{\epsilon}2^{|\lambda|})\delta_{\lambda,\mu}\right)_{\lambda,\mu\in\mathcal{J}}$. *Then whenever (16) holds with* $\gamma > 1$ *one has for every* $\mathbf{v} \in \ell_2$

$$\left(2(c_0^{-2}+c_1^{-1})\right)^{-1/2}\|\mathbf{v}\|_{\ell_2} \le \|\mathbf{v}^T\mathbf{D}_\epsilon^{-1}\Psi\|_{\mathcal{H}_\epsilon} \le \left(C_0^2+C_1^2\right)^{1/2}\|\mathbf{v}\|_{\ell_2}. \tag{17}$$

Proof: Let $v = \boldsymbol{d}^T\Psi \;\; (= (\mathbf{D}_\epsilon\boldsymbol{d})^T\mathbf{D}_\epsilon^{-1}\Psi)$. We wish to show that $\|\mathbf{D}_\epsilon\boldsymbol{d}\|_{\ell_2} \sim \|v\|_{\mathcal{H}_\epsilon}$):

$$\begin{aligned}\|\{(1+\sqrt{\epsilon}2^{|\lambda|})d_\lambda\}_{\lambda\in\mathcal{J}}\|^2_{\mathcal{J}} &\le 2\sum_{\lambda\in\mathcal{J}}\left(|d_\lambda|^2+\epsilon 2^{2|\lambda|}|d_\lambda|^2\right)\\ &\overset{(NE)}{\le} 2\left(c_0^{-2}+c_1^{-2}\right)\left\{\|v\|^2_{L_2}+\epsilon|v|^2_{H^1}\right\}\\ &= 2\left(c_0^{-2}+c_1^{-2}\right)\|v\|^2_{\mathcal{H}_\epsilon}\end{aligned}$$

Conversely one has

$$\begin{aligned}\|v\|^2_{L_2}+\epsilon|v|^2_{H^1} &\overset{(NE)}{\le} C_0^2\|\mathbf{d}\|^2_{\ell_2}+\epsilon C_1^2\|\mathbf{D}^1\mathbf{d}\|^2_{\ell_2} \le (C_0^2+C_1^2)\sum_{\lambda\in\mathcal{J}}(1+\epsilon 2^{2|\lambda|})|d_\lambda|^2\\ &\le (C_0^2+C_1^2)\sum_{\lambda\in\mathcal{J}}(1+\sqrt{\epsilon}2^{|\lambda|})^2|d_\lambda|^2,\end{aligned}$$

which finishes the proof. □

The next consequence is of general nature and concerns *duality.*

Remark 2 *Let* $\mathcal{H}$ *be a Hilbert space,* $\langle\cdot,\cdot\rangle : \mathcal{H}\times\mathcal{H}' \to \mathbb{R}$, *and suppose that*

$$c\|\mathbf{v}\|_{\ell_2} \le \|\mathbf{v}^T\Theta\|_{\mathcal{H}} \le C\|\mathbf{v}\|_{\ell_2}, \tag{18}$$

i.e., Θ *is a Riesz-basis for* $\mathcal{H}$. *Then one has*

$$C^{-1}\|\langle\Theta, v\rangle\|_{\ell_2} \le \|v\|_{\mathcal{H}'} \le c^{-1}\|\langle\Theta, v\rangle\|_{\ell_2}. \tag{19}$$

An important Application of (19) is the case $\mathcal{H} = H_0^1(\Omega)$, $\Theta := \mathbf{D}^{-1}\Psi$ which gives

$$C_1^{-1}\|\mathbf{D}^{-1}\langle\Psi,w\rangle\|_{\ell_2} \leq \|w\|_{H^{-1}(\Omega)} \leq c_1^{-1}\|\mathbf{D}^{-1}\langle\Psi,w\rangle\|_{\ell_2}. \qquad (20)$$

The H^{-1} norm arise naturally in connection with second order elliptic equations. Its evaluation is, for instance, important in connection with least squares formulations [17, 18] and poses serious difficulties in the finite element context.

Proof of Remark 2: We proceed in three steps.

1) Consider the mapping $F : \ell_2 \to \mathcal{H}$, $F : \mathbf{v} \to \mathbf{v}^T\Theta$. Then (18) means that

$$\|F\|_{\ell_2\to\mathcal{H}} = C, \quad \|F^{-1}\|_{\mathcal{H}\to\ell_2} = c^{-1}. \qquad (21)$$

2) For the adjoint $F^* : \mathcal{H}' \to \ell_2$, defined by $\langle F\mathbf{v}, w\rangle = \mathbf{v}^T F^* w$, one then has on one hand

$$\begin{aligned}\|F^*w\|_{\ell_2} &= \sup_{\mathbf{v}} \frac{(F^*w)^T\mathbf{v}}{\|\mathbf{v}\|_{\ell_2}} = \sup_{\mathbf{v}} \frac{\langle F\mathbf{v}, w\rangle}{\|\mathbf{v}\|_{\ell_2}} \\ &\leq \sup_{\mathbf{v}} \frac{\|F\mathbf{v}\|_{\mathcal{H}}\|w\|_{\mathcal{H}'}}{\|\mathbf{v}\|_{\ell_2}} = \|F\|_{\ell_2\to\mathcal{H}}\|w\|_{\mathcal{H}'},\end{aligned}$$

and on the other hand,

$$\begin{aligned}\|F\mathbf{v}\|_{\mathcal{H}} &= \sup_{w} \frac{\langle F\mathbf{v}, w\rangle}{\|w\|_{\mathcal{H}'}} \sup_{w} \frac{(F^*w)^T\mathbf{v}}{\|w\|_{\mathcal{H}'}} \\ &\leq \sup_{w} \frac{\|F^*w\|_{\ell_2}\|\mathbf{v}\|_{\ell_2}}{\|w\|_{\mathcal{H}'}} = \|F^*\|_{\mathcal{H}'\to\ell_2}\|\mathbf{v}\|_{\ell_2}.\end{aligned}$$

Thus we conclude that

$$\|F\|_{\ell_2\to\mathcal{H}} = \|F^*\|_{\mathcal{H}'\to\ell_2} = C, \quad \|F^{-1}\|_{\mathcal{H}\to\ell_2} = \|(F^*)^{-1}\|_{\ell_2\to\mathcal{H}'} = c^{-1}. \qquad (22)$$

3) It remains to identify F^*w by applying it to the sequence $\mathbf{e}_\lambda := (\delta_{\lambda,\mu} : \mu \in \mathcal{J})$. In fact, $(F^*w)_\lambda = (F^*w)^T\mathbf{e}_\lambda = \langle F\mathbf{e}_\lambda, w\rangle = \langle\theta_\lambda, w\rangle$ which means

$$F^*w = \langle\Theta, w\rangle,$$

whence the assertion follows. □

Riesz-Bases – Biorthogonality: We pause to stress the role of biorthogonality in the context of Riesz bases. Recall that under the assumption (18) the mappings

$$F : \ell_2 \to \mathcal{H} \;\; F : \mathbf{v} \to \mathbf{v}^T\Theta, \quad F^* : \mathcal{H}' \to \ell_2 \quad \langle F\mathbf{v}, w\rangle = \mathbf{v}^T F^* w,$$

are *topological isomorphisms* and that $F\mathbf{e}_\lambda = \theta_\lambda$. Defining the collection $\tilde{\Theta}$ by $\tilde{\theta}_\lambda := (F^*)^{-1}\mathbf{e}_\lambda$, we obtain

$$\langle \theta_\lambda, \tilde{\theta}_\mu \rangle = \langle F\mathbf{e}_\lambda, (F^*)^{-1}\mathbf{e}_\mu \rangle = \mathbf{e}_\lambda^T \mathbf{e}_\mu = \delta_{\lambda,\mu},$$

which means

$$\langle \Theta, \tilde{\Theta} \rangle = \mathbf{I}. \tag{23}$$

Thus one has $w = \langle w, \Theta \rangle \tilde{\Theta}$ and (19) simply means that $\tilde{\Theta}$ is a Riesz basis for $\mathcal{H}'$. Hence a Riesz basis for $\mathcal{H}$ always entails a Riesz basis for $\mathcal{H}'$. Of course, $\Theta = \tilde{\Theta}$ when Θ is an orthonormal basis and $\mathcal{H}$ is identified with $\mathcal{H}'$. The following consequence often guides the construction of wavelet bases satisfying (16) also for $s = 0$.

Corollary 3 *When $\mathcal{H} = L_2(\Omega) = \mathcal{H}'$ for every Riesz basis Ψ of L_2 there exists a biorthogonal basis $\tilde{\Psi}$ which is also a Riesz basis for L_2.*

We conclude this section with some remarks on

Further norm equivalences - Besov-Spaces: Norm equivalence of the type (16) extend to other function spaces such as Besov spaces. For a more detailed treatment the reader is referred, for instance, to [11, 24, 51, 53, 52], see also Section 10 for definitions.

Note first that (16) can be viewed for $0 < p < \infty$, $0 < q \leq \infty$ as a special case of

$$\|v\|^q_{B^t_q(L_p)} \sim \|v\|^q_{L_p} + \sum_{\lambda \in \mathcal{J}} 2^{tq|\lambda|} \|\langle v, \tilde{\psi}_\lambda \rangle \psi_\lambda\|^q_{L_p}, \tag{24}$$

which holds again for some range of $t > 0$ depending on the regularity of the wavelets. Now, renormalizing

$$\psi_{\lambda,p} := 2^{d|\lambda|(\frac{1}{p}-\frac{1}{2})}\psi_\lambda \quad \text{so that} \quad \|\psi_{\lambda,p}\|_{L_p} \sim 1,$$

we denote by $\Psi_p := \{\psi_{\lambda,p} : \lambda \in \mathcal{J}\}$ the L_p-normalized version of Ψ and note that one still has a biorthogonal pair $\langle \Psi_p, \tilde{\Psi}_{p'} \rangle = \mathbf{I}$. Moreover, whenever we have an embedding $B^t_q(L_\tau) \subset L_p$, (24) can be restated as

$$\|v\|^q_{B^t_q(L_\tau)} \sim \|v\|^q_{L_\tau} + \sum_{j=0}^{\infty} \left(2^{jd(\frac{t}{d}+\frac{1}{p}-\frac{1}{\tau})} \|\langle \tilde{\Psi}_{j,p'}, v \rangle\|_{\ell_\tau} \right)^q, \tag{25}$$

where $\tilde{\Psi}_{j,p'} := \{\tilde{\psi}_{\lambda,p'} : |\lambda| = j\}$. Note that the embedding $B^t_q(L_\tau) \subset L_p$ holds, by the Sobolev embedding theorem, as long as

$$\frac{1}{\tau} \leq \frac{t}{d} + \frac{1}{p}, \tag{26}$$

(with a restricted range for q when equality holds in (26)). In particular, in the extreme case one has

$$\frac{t}{d} + \frac{1}{p} = \frac{1}{\tau} \quad \rightsquigarrow \quad \|v\|_{B^t_\tau(L_\tau)} \sim \|v\|_{L_\tau} + \|\langle \tilde{\Psi}_{p'}, v \rangle\|_{\ell_\tau}, \tag{27}$$

i.e., the Besov norm is equivalent to an ℓ_τ-norm, where no scale dependent weight arises any more. This indicates that the embedding is not compact. It is however, an important case that plays a central role in nonlinear approximation and adaptivity as explained later.

Figure 4 illustrates the situation. With every point in the $(1/p, s)$ coordinate plane we associate (a collection of) spaces with smoothness s in L_p. Left of the critical line through $(1/p, 0)$ with slope d all spaces are still embedded in L_p (with a compact embedding strictly left of the critical line), see [24, 51].

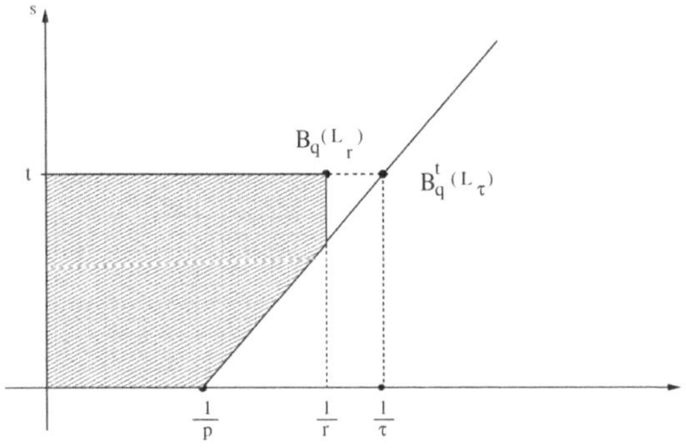

Fig. 4. Critical embedding line

4 Criteria for (NE)

In view of the importance of (NE), we discuss briefly ways of affirming its validity. Fourier techniques are typically applied when dealing with bases living on the torus or in Euclidean space. Different concepts are needed when dealing with more general domain geometries. We have already seen that the existence of a biorthogonal Riesz basis comes with a Riesz basis. Thus biorthogonality is a *necessary* but unfortunately not sufficient criterion.

4.1 What Could Additional Conditions Look Like?

Let us see next what kind of conditions, in addition to biorthogonality, are related to (NE). To this end, suppose that (NE) holds for $0 \le t < \gamma$, so that

$$\|\sum_{|\lambda|<J} d_\lambda \psi_\lambda\|_{H^t} \overset{(NE)}{\sim} \|\{2^{t|\lambda|} d_\lambda\}_{|\lambda|<J}\|_{\ell_2} \lesssim 2^{Jt} \|\{d_\lambda\}_{|\lambda|<J}\|_{\ell_2}$$
$$\overset{(NE)}{\sim} 2^{Jt} \|\sum_{|\lambda|<J} d_\lambda \psi_\lambda\|_{L_2}.$$

Thus defining the hierarchy of spaces

$$S_j := \mathrm{clos}_{H^t}\mathrm{span}\,(\psi_\lambda : |\lambda| < j), \tag{28}$$

we see that (NE) implies the

Inverse (or Bernstein) Estimate:

$$\|v\|_{H^t} \lesssim 2^{jt}\|v\|_{L_2}, \quad v \in S_j,\ t < \gamma. \tag{29}$$

Furthermore, consider the *canonical projectors*

$$Q_j v := \sum_{|\lambda|<j} \langle v, \tilde{\psi}_\lambda\rangle \psi_\lambda, \quad Q_j^* v := \sum_{|\lambda|<j} \langle v, \psi_\lambda\rangle \tilde{\psi}_\lambda, \tag{30}$$

and note that

$$\begin{aligned} \|v - Q_j v\|_{L_2} &\lesssim \left(\sum_{|\lambda|\geq j} |\langle v, \tilde{\psi}_\lambda\rangle|^2\right)^{1/2} \leq 2^{-jt}\left(\sum_{|\lambda|\geq j} 2^{2t|\lambda|}|\langle v, \tilde{\psi}_\lambda\rangle|^2\right)^{1/2} \\ &\overset{(\mathrm{NE})}{\lesssim} 2^{-jt}\|v\|_{H^t}. \end{aligned}$$

This entails the

Direct (or Jackson) Estimate:

$$\inf_{v_j \in S_j} \|v - v_j\|_{L_2} \lesssim 2^{-jt}\|v\|_{H^t}, \quad t < \gamma. \tag{31}$$

Conversely, it will be seen below that direct and inverse estimates imply the validity of (NE) for a certain range of t.

4.2 Fourier- and Basis-free Criteria

Note that the above consequences of (NE) are actually properties of the multiresolution spaces S_j, not of the specific bases. Let us therefore use possibly basis-free formulations. To this end, note that details between two successive spaces S_j and S_{j+1} can be expressed as

$$\sum_{|\lambda|=j} \langle v, \tilde{\psi}_\lambda\rangle \psi_\lambda = (Q_{j+1} - Q_j)v.$$

Collections of functions that are stable on each level are relatively easy to construct. In fact, when the biorthogonal basis is also local, this is easy to see, as explained later, see Remark 5 in Section 5.1. In such a case one has

$$\|(Q_{j+1} - Q_j)v\|_{L_2}^2 \sim \sum_{|\lambda|=j} |\langle v, \tilde{\psi}_\lambda\rangle|^2,$$

which means

$$N_{s,\mathcal{Q}}(v) := \left(\sum_{j=0}^{\infty} 2^{j2s} \|(Q_{j+1} - Q_j)v\|_{L_2}^2 \right)^{1/2} \sim \left(\sum_{\lambda \in \mathcal{J}} 2^{2s|\lambda|} |\langle v, \tilde{\psi}_\lambda \rangle|^2 \right)^{1/2}.$$

Moreover, biorthogonality of $\Psi, \tilde{\Psi}$ means that the operators Q_j defined by (30) *commute*. Thus, the seemingly relevant properties like direct and inverse estimates as well as biorthogonality can be formulated without explicit reference to the specific bases $\Psi, \tilde{\Psi}$ (which will be important for construction principles). In fact, in summary, one is led by the above discussion to ask the following question:

- Given a *multiresolution sequence* of nested spaces $\mathcal{S} = \{S_j\}_{j \in \mathbb{N}_0}$: $S_j \subset H^s$ for $s < \gamma$, and an associate sequence $\mathcal{Q} = \{Q_j\}_{j \in \mathbb{N}_0}$ of uniformly L_2-bounded projectors mapping L_2 onto S_j, such that
- *the commutator property (C):*

$$Q_l Q_j = Q_l, \quad l \leq j, \tag{32}$$

- *the Jackson estimate (J):*

$$\inf_{v_j \in V_j} \|v - v_j\|_{L_2} \lesssim 2^{-m'j} \|v\|_{H^{m'}}, \quad v \in H^{m'} \tag{33}$$

  and the
- *Bernstein estimate (B):*

$$\|v_j\|_{H^s} \lesssim 2^{sj} \|v_j\|_{L_2}, \quad v_j \in V_j,\ s < \gamma', \tag{34}$$

  hold for $S_j = V_j$ and some $\gamma' > 0, m' \in \mathbb{N}$,

can one ensure that $N_{s,\mathcal{Q}}(\cdot) \sim \|\cdot\|_{H^s}$?

The following statement is a special case of a more general result from [39].

Theorem 1. *For $\mathcal{S}$, $\mathcal{Q}$ as above suppose that (32) and (34), (33) hold for $V_j = S_j$ with $m' = m > \gamma' = \gamma > 0$. Then*

$$\|v\|_{H^s} \sim \left(\sum_{j=0}^{\infty} 2^{2sj} \|(Q_j - Q_{j-1})v\|_{L_2}^2 \right)^{1/2}, \quad 0 < s < \gamma. \tag{35}$$

Moreover, if (33) and (34) also hold for $V_j = \tilde{S}_j := \operatorname{range} Q_j^*$ with $m' = \tilde{m} > \gamma' = \tilde{\gamma} > 0$, then the above equivalence (35) also holds for $-\tilde{\gamma} < s < \gamma$, where for $s < 0$ it is understood that $H^s = (H^{-s})'$.

A few comments are in order. Note that (35) is easier to realize for $s > 0$. The case $s \leq 0$ requires more effort and involves the *dual multiresolution sequence* $\tilde{\mathcal{S}}$ as well. The space H^s above may have incorporated homogeneous boundary conditions.

As for the usefulness of the above criterion for the construction of a Riesz basis for L_2, the following remarks should be kept in mind. One actually starts usually with a multiresolution sequence $\mathcal{S}$ where the S_j are then *not* given as spans of wavelets but of *single scale bases* Φ_j. They consist of compactly supported functions like scaling functions in classical settings or by standard finite element nodal basis functions, see the examples in Section 2. One then tries to construct suitable *complement bases* spanning complements between any two successive spaces in $\mathcal{S}$. It is of course not clear how to get a good guess for such complement bases whose union would be a candidate for Ψ. The difficulty is to choose these complements in such way that they give rise to a dual multiresolution (which is completely determined once the ψ_λ are chosen) satisfying the conditions (J), (B) as well for some range. It will be seen later how to solve this problem in a systematic fashion, provided that for a *single scale basis* Φ_j for S_j biorthogonal bases $\tilde{\Phi}_j$ can be found which then define $\tilde{\mathcal{S}}$. One can then construct Ψ and $\tilde{\Psi}$ based only on the knowledge of the collections Φ_j, $\tilde{\Phi}_j$.

Although the construction of dual single scale pairs Φ_j, $\tilde{\Phi}_j$ is possible in some important cases, one faces serious difficulties with this approach when dealing with finite elements on nested triangulations. Therefore criteria, that do not require explicit knowledge of dual pairs Φ_j, $\tilde{\Phi}_j$, are desirable. We outline next such criteria following essentially the developments in [49]. The key is that both multiresolution sequences $\mathcal{S}$ and $\mathcal{S}$ are *prescribed.* It will be seen that one need not know corresponding biorthogonal single scale bases to start with, but it rather suffices to ensure that single scale bases for these spaces can be *biorthogonalized.* This latter property can be expressed as follows.

Remark 4 *Let $\mathcal{S}$ and $\tilde{\mathcal{S}}$ be* two given *multi-resolution sequences satisfying* $\dim S_j = \dim \tilde{S}_j$ *and*

$$\inf_{v_j \in S_j} \sup_{\tilde{v}_j \in \tilde{S}_j} \frac{|\langle v_j, \tilde{v}_j \rangle|}{\|v_j\|_{L_2(\Omega)} \|\tilde{v}_j\|_{L_2(\Omega)}} \gtrsim 1. \tag{36}$$

Then there exists a sequence $\mathcal{Q}$ of uniformly L_2-bounded projectors with ranges $\mathcal{S}$ such that

$$\mathrm{range}\,(id - Q_j) = (\tilde{S}_j)^{\perp_{L_2}}, \quad \mathrm{range}\,(id - Q_j^*) = (S_j)^{\perp_{L_2}}. \tag{37}$$

Moreover, $\mathcal{Q}$ satisfies the commutator property (C), see (32).

This leads to a version of Theorem 1, that *prescribes* both multiresolution sequences $\mathcal{S}$ and $\tilde{\mathcal{S}}$ without requiring explicit knowledge neither of the dual generator bases nor of the dual wavelets.

Theorem 2. *If in addition to the hypotheses of Remark 4 $\mathcal{S}$ and $\tilde{\mathcal{S}}$ satisfy direct and inverse estimates (33) and (34) with respective parameters $m, \tilde{m} \in \mathbb{N}$ and $\gamma, \tilde{\gamma} > 0$, then one has for any $w_j \in \operatorname{range}(Q_j - Q_{j-1})$, Q_j given by Remark 4,*

$$\left\| \sum_{j=0}^{\infty} w_j \right\|_{H^s}^2 \lesssim \sum_{j=0}^{\infty} 2^{2sj} \|w_j\|_{L_2}^2, \quad s \in [-\tilde{m}, \gamma), \tag{38}$$

and

$$\sum_{j=0}^{\infty} 2^{2sj} \|(Q_j - Q_{j-1})v\|_{L_2}^2 \lesssim \|v\|_{H^s}^2, \quad s \in (-\tilde{\gamma}, m]. \tag{39}$$

Thus for $s \in (-\tilde{\gamma}, \gamma)$ (38), (39), $v \to \{(Q_j - Q_{j-1})v\}_j$ is a bounded mapping from H^s onto $\ell_{2,s}(\mathcal{Q}) := \{\{w_j\}_j : w_j \in \operatorname{range}(Q_j - Q_{j-1}), \|\{w_j\}_j\|_{\ell_{2,s}(\mathcal{Q})} := \left(\sum_{j=0}^{\infty} 2^{2sj} \|w_j\|_{L_2(\Omega)}^2\right)^{1/2} < \infty\}$, with bounded inverse $\{w_j\}_j \to \sum_{j=0}^{\infty} w_j$, i.e., for $s \in (-\tilde{\gamma}, \gamma)$ the above relations hold with '$\lesssim$' replaced by '$\sim$'.

Clearly Theorem 2 implies Theorem 1. We sketch now the main steps in the proof of Theorem 2 (see [39, 49] for more details). Define orthogonal projectors $P_j : L_2 \to \tilde{S}_j$ by

$$\langle v, \tilde{v}_j \rangle = \langle P_j v, \tilde{v}_j \rangle, \quad v \in L_2, \ \tilde{v}_j \in \tilde{S}_j,$$

and set $R_j : P_j|_{S_j}$ so that $\|R_j\|_{L_2} \leq 1$. Then (36) implies $\|R_j v_j\|_{L_2} \gtrsim \|v_j\|_{L_2}$, $v_j \in S_j$. Now we claim that range $R_j = \tilde{S}_j$, since otherwise there would exist $\tilde{v}'_j \in \tilde{S}_j$, $\tilde{v}'_j \perp_{L_2}$ range R_j, contradicting (36). Thus, the $R_j^{-1} : \tilde{S}_j \to S_j$ are uniformly L_2-bounded. Then the projectors $Q_j := R_j^{-1} Q_j : L_2 \to S_j$ are uniformly L_2 bounded, range $Q_j = S_j$ and $\langle Q_j v, \tilde{v}_j \rangle = \langle v, \tilde{v}_j \rangle$, which provides range $(I - Q_j) \subset (\tilde{S}_j)^{\perp_{L_2}}$.

Conversely, $v \in (\tilde{S}_j)^{\perp_{L_2}}$ implies range $(I - Q_j) \subset (\tilde{S}_j)^{\perp_{L_2}}$, which confirms analogous properties for the adjoints Q_j^*.

Note that $\tilde{S}_j \subset \tilde{S}_{j+1}$, which implies (C) $Q_j Q_{j+1} = Q_j$. □

As for Theorem 2, observe first

$$\|w_j\|_{H^{s\pm\epsilon}} \lesssim 2^{j(s\pm\epsilon)} \|w_j\|_{L_2}, \quad \forall\, w_j \in \operatorname{range}(Q_j - Q_{j-1}), \ s \pm \epsilon \in [-\tilde{m}, \gamma). \tag{40}$$

In fact, (40) follows from (B) (34) for $s \pm \epsilon \in [0, \gamma)$. For $t := s \pm \epsilon \in [-\tilde{m}, 0]$ and any $w_j \in \operatorname{range}(Q_j - Q_{j-1})$, one has

$$\begin{aligned} \|w_j\|_{H^t} &= \sup_{z \in H^{-t}} \frac{\langle w_j, z \rangle}{\|z\|_{H^{-t}}} = \sup_{z \in H^{-t}} \frac{\langle w_j, (Q_j^* - Q_{j-1}^*) z \rangle}{\|z\|_{H^{-t}}} \\ &\lesssim \sup_{z \in H^{-t}} \frac{\|w_j\|_{L_2} \inf_{\tilde{v}_{j-1} \in \tilde{S}_{j-1}} \|z - \tilde{v}_{j-1}\|_{L_2}}{\|z\|_{H^{-t}}} \overset{(J)}{\lesssim} 2^{tj} \|w_j\|_{L_2}. \end{aligned}$$

Hence

$$\Big\|\sum_j w_j\Big\|_{H^s}^2 = \Big\langle \sum_j w_j, \sum_l w_l \Big\rangle_{H^s} \lesssim \sum_j \sum_{l \geq j} \|w_j\|_{H^{s+\epsilon}} \|w_l\|_{H^{s-\epsilon}}$$

$$\overset{(B)}{\lesssim} \sum_j \sum_{l \geq j} 2^{\epsilon j} 2^{-l\epsilon} (2^{sk} \|w_j\|_{L_2})(2^{sl} \|w_l\|_{L_2}) \lesssim \sum_j 2^{2sj} \|w_j\|_{L_2}^2.$$

Thus, we have shown

$$\|v\|_{H^s} \lesssim N_{s,\mathcal{Q}}(v), \quad s \in [-\tilde{m}, \gamma), \tag{41}$$

which is (38). Interchanging the roles of $\mathcal{S}$ and $\tilde{\mathcal{S}}$, the same argument gives

$$\|v\|_{H^s} \lesssim N_{s,\mathcal{Q}^*}(v), \quad s \in [-m, \tilde{\gamma}). \tag{42}$$

In order to prove now (39), note first that

$$\begin{aligned} N_{s,\mathcal{Q}}(v)^2 &= \sum_j 2^{2sj} \langle (Q_j - Q_{j-1})v, (Q_j - Q_{j-1})v \rangle \\ &= \Big\langle \sum_j 2^{2sj} (Q_j^* - Q_{j-1}^*)(Q_j - Q_{j-1})v, v \Big\rangle \\ &\leq \| \underbrace{\sum_j 2^{2sj} (Q_j^* - Q_{j-1}^*)(Q_j - Q_{j-1})v}_{:=w} \|_{H^{-s}} \|v\|_{H^s} \\ &\overset{(42)}{\lesssim} N_{-s,\mathcal{Q}^*}(w) \|v\|_{H^s}, \quad s \in (-\tilde{\gamma}, m] \end{aligned} \tag{43}$$

Since by (C) (see (32)) $(Q_l^* - Q_{l-1}^*)(Q_j^* - Q_{j-1}^*) = \delta_{j,l}(Q_l^* - Q_{l-1}^*)$ one has

$$\|(Q_l^* - Q_{l-1}^*)w\|_{L_2} = 2^{2sl} \|(Q_l^* - Q_{l-1}^*)(Q_l - Q_{l-1})v\|_{L_2} \lesssim 2^{2sl} \|(Q_l - Q_{l-1})v\|_{L_2}.$$

This gives

$$\begin{aligned} N_{-s,\mathcal{Q}^*}(w) &= \Big(\sum_l 2^{-2sl} \|(Q_l^* - Q_{l-1}^*)w\|_{L_2}^2 \Big)^{1/2} \\ &\lesssim \Big(\sum_j 2^{-2sl} 2^{4sl} \|(Q_l - Q_{l-1})v\|_{L_2}^2 \Big)^{1/2} = N_{s,\mathcal{Q}}(v). \end{aligned}$$

Therefore we can bound $N_{-s,\mathcal{Q}^*}(w)$ on the right hand side of (43) by a constant multiple of $N_{s,\mathcal{Q}}(v)$. Dividing both sides of the resulting inequality by $N_{s,\mathcal{Q}}(v)$, yields (39). □

Direct and inverse estimates (J), (B) are satisfied for all standard hierarchies of trial spaces where m is the order of polynomial exactness.

- A possible strategy for (J) is to construct L_2-bounded local projectors onto S_j and use the reproduction of polynomials and corresponding local polynomial inequalities;
- (B) follows from stability and rescaling arguments. The simplest case is $s \in \mathbb{N}$. One first establishes an estimate on a reference domain, then rescales and sums up the local quantities.

5 Multiscale Decompositions – Construction and Analysis Principles

We indicate next some construction principles that allow us to make use of the above stability criteria. The main tool is the notion of multiresolution hierarchies.

5.1 Multiresolution

The common starting point for the construction of multiscale bases is an ascending sequence of spaces

$$S_0 \subset S_1 \subset S_2 \subset \ldots L_2(\Omega), \qquad \overline{\bigcup_j S_j} = L_2(\Omega),$$

which are spanned by *single scale bases*

$$S_j = \operatorname{span} \Phi_j \; =: S(\Phi_j), \quad \Phi_j = \{\phi_{j,k} : k \in \mathcal{I}_j\}.$$

One then seeks decompositions

$$S_{j+1} = S_j \bigoplus W_j$$

along with corresponding *complement bases*

$$W_j = \operatorname{span} \Psi_j, \quad \Psi_j = \{\psi_\lambda : \lambda \in \mathcal{J}_j\}.$$

The union of the coarse scale basis Φ_0 and all the complement bases provides a *multi-scale basis*

$$\Psi := \bigcup_{j \in \mathbb{N}_0} \Psi_j \qquad (\Psi_{-1} := \Phi_0),$$

which is a candidate for a wavelet basis.

Before discussing how to find *suitable* complement bases, it is important to distinguish several stability notions.

Uniform single scale stability: This means that the relations

$$\|\mathbf{c}\|_{l_2(\mathcal{I}_j)} \sim \|\mathbf{c}^T \Phi_j\|_{L_2}, \qquad \|\boldsymbol{d}\|_{l_2(\mathcal{J}_j)} \sim \|\boldsymbol{d}^T \Psi_j\|_{L_2} \tag{44}$$

hold uniformly in j. Single scale stability is easily established, for instance, when *local dual bases* are available.

Remark 5 *Suppose $\Phi_j, \tilde{\Phi}_j$ are* dual pairs *of single scale bases, i.e.,*

$$\langle \Phi_j, \tilde{\Phi}_j \rangle = \mathbf{I}, \quad \|\phi_{j,k}\|_{L_2} \sim 1, \quad \|\tilde{\phi}_{j,k}\|_{L_2} \sim 1,$$

$$\operatorname{diam}(\operatorname{supp} \Phi_j), \ \operatorname{diam}(\operatorname{supp} \tilde{\Phi}_j) \sim 1.$$

Then (44) holds for Φ_j and $\tilde{\Phi}_j$.

Proof: Let $\sigma_{j,k}$ denote the support of $\phi_{j,k}$ and $\Box_{j,k} := 2^{-j}(k + [0,1]^d)$. Then for $v = \mathbf{c}^T \Phi_j$ one has

$$\|\mathbf{c}\|_{\ell_2}^2 = \|\langle v, \tilde{\Phi}_j \rangle\|_{\ell_2}^2 \ \lesssim \ \sum_{k \in \mathcal{I}_j} \|v\|_{L_2(\tilde{\sigma}_{j,k})}^2 \ \lesssim \ \|v\|_{L_2}^2.$$

Conversely,

$$\|v\|_{L_2(\Box_{j,k})}^2 \leq \left(\sum_{\sigma_{j,m} \cap \Box_{j,k} \neq \emptyset} |c_m| \|\phi_{j,m}\|_{L_2(\Box_{j,k})} \right)^2 \ \lesssim \ \sum_{\sigma_{j,m} \cap \Box_{j,k} \neq \emptyset} |c_m|^2,$$

which, upon summation, yields

$$\|v\|_{L_2}^2 \ \lesssim \ \|\mathbf{c}\|_{\ell_2}^2.$$

This finishes the proof. □

Of course, this does not mean that the multiscale basis Ψ is stable in the sense that $\|\boldsymbol{d}\|_{\ell_2} \sim \|\boldsymbol{d}^T \Psi\|_{L_2}$, which we refer to as *stability over all levels*, or equivalently the Riesz-basis property in L_2.

5.2 Stability of Multiscale Transformations

Let us pause here to indicate a first instance where the Riesz-basis property has computational significance. To this end, note that for each j the elements of S_j have two equivalent representations

$$\sum_{l=-1}^{j-1} \sum_{\lambda \in \mathcal{J}_l} d_\lambda \psi_\lambda = \sum_{k \in \mathcal{I}_j} c_k \phi_{j,k}$$

$$\mathbf{T}_j \ : \ \boldsymbol{d} \mapsto \ \mathbf{c}$$

and the corresponding arrays of single-scale, respectively multiscale coefficients $\mathbf{c}$, $\boldsymbol{d}$ are interrelated by the *multiscale transformation* $\mathbf{T}_j$, see Section 2.1.

Remark 6 (cf. [23, 39]) *Assume that the Φ_j are uniformly L_2-stable. Then*

$$\|\mathbf{T}_j\|, \ \|\mathbf{T}_j^{-1}\| = \mathcal{O}(1) \quad \Longleftrightarrow \quad \Psi \ \text{ is a Riesz basis in } \ L_2.$$

Proof: Let $\Psi^j = \{\psi_\lambda : |\lambda| < j\}$ and $v = \boldsymbol{d}_j^T \Psi^j = \mathbf{c}_j^T \Phi_j$. Assume that Ψ is a Riesz basis. Then one has

$$\|\mathbf{c}_j\|_{\ell_2} \sim \|v\|_{L_2} = \|\boldsymbol{d}_j^T \Psi^j\|_{L_2} \sim \|\boldsymbol{d}_j\|_{\ell_2} = \|\mathbf{T}_j^{-1}\mathbf{c}_j\|_{\ell_2}.$$

Conversely:

$$\|\boldsymbol{d}_j\|_{\ell_2} = \|\mathbf{T}_j^{-1}\mathbf{c}_j\|_{\ell_2} \sim \|\mathbf{c}_j\|_{\ell_2} \sim \|v\|_{L_2},$$

whence the assertion follows. □

Stability over all scales is a more involved issue and will be addressed next.

5.3 Construction of Biorthogonal Bases – Stable Completions

We have seen in Section 4 that biorthogonality is necessary for the Riesz basis property. Once a pair of biorthogonal bases is given, one can use Theorem 1 or Theorem 2 to check the desired norm equivalences (16), in particular, for $s = 0$.

The construction and analysis of biorthogonal wavelets on $\mathbb{R}$ or $\mathbb{R}^d$ is greatly facilitated by Fourier methods (see [25, 50]). For more general domains the construction of *good complement spaces* W_j is based on different strategies. A rough road map looks as follows:

- Construct stable dual multiresolution sequences

  $$\mathcal{S} = \{S_j\}_{j\in\mathbb{N}_0}, \quad S_j = \operatorname{span}(\Phi_j), \quad \tilde{\mathcal{S}} = \{\tilde{S}_j\}_{j\in\mathbb{N}_0}, \quad \tilde{S}_j = \operatorname{span}(\tilde{\Phi}_j),$$

  such that $\langle \Phi_j, \tilde{\Phi}_j\rangle = \mathbf{I}$, $j \in \mathbb{N}_0$.
- Construct some simple *initial* complement spaces $\check{W}_j$.
- *Change* the initial complements into better ones W_j.

We will comment next on these steps.

5.4 Refinement Relations

Stability (44) and nestedness of the spaces S_j imply *refinement equations* of the form

$$\phi_{j,k} = \sum_{l\in\mathcal{I}_{j+1}} m_{j,l,k}\phi_{j+1,l}, \tag{45}$$

which we briefly express as

$$\Phi_j^T = \Phi_{j+1}^T \mathbf{M}_{j,0}. \tag{46}$$

Here the columns of the refinement matrix $\mathbf{M}_{j,0}$ consist of the arrays of *mask coefficients* $m_{j,l,k}$. The objective is then to find a basis $\Psi_{j+1} = \{\psi_\lambda : |\lambda| = j\} \subset S_{j+1}$, i.e.,

$$\Psi_{j+1}^T = \Phi_{j+1}^T \mathbf{M}_{j,1}, \tag{47}$$

that spans a complement of S_j in S_{j+1}. It is easy to see that this is equivalent to the fact that the matrix $\mathbf{M}_j := (\mathbf{M}_{j,0}, \mathbf{M}_{j,1})$ is *invertible*, i.e, for $\mathbf{M}^{-1} =: \mathbf{G}_j = \binom{\mathbf{G}_{j,0}}{\mathbf{G}_{j,1}}$ one has

$$\Phi_{j+1}^T = \Phi_j^T \mathbf{G}_{j,0} + \Psi_{j+1}^T \mathbf{G}_{j,1} \iff \mathbf{M}_j \mathbf{G}_j = \mathbf{G}_j \mathbf{M}_j = \mathbf{I}. \tag{48}$$

Remark 7 [23] *The bases Ψ_j, Φ_j are uniformly stable in the sense of (44) if and only if*

$$\|\mathbf{M}_j\|_{\ell_2 \to \ell_2}, \|\mathbf{G}_j\|_{\ell_2 \to \ell_2} = \mathcal{O}(1). \tag{49}$$

The matrix $\mathbf{M}_{j,1}$ is called a *completion* of the refinement matrix $\mathbf{M}_{j,0}$. The sequence $\{\mathbf{M}_{j,1}\}_j$ is called (uniformly) *stable completions* when (49) holds.

Let us illustrate these notions by the hierarchical complement bases from Section 2.3. From the refinement equations one readily infers (see Figure 5)

$$\mathbf{M}_{j,0} = \begin{pmatrix} \frac{1}{2\sqrt{2}} & 0 & 0 & \dots & \\ \frac{1}{\sqrt{2}} & 0 & 0 & \dots & \\ \frac{1}{2\sqrt{2}} & \frac{1}{2\sqrt{2}} & 0 & \dots & \\ 0 & \frac{1}{\sqrt{2}} & 0 & \dots & \\ 0 & \frac{1}{2\sqrt{2}} & \frac{1}{2\sqrt{2}} & 0 & \dots \\ 0 & 0 & \dots & & \\ \vdots & \vdots & \vdots & & \vdots \\ & & & & 0 \\ & & & \frac{1}{2\sqrt{2}} & \frac{1}{2\sqrt{2}} \\ 0 & \dots & & 0 & \frac{1}{\sqrt{2}} \\ 0 & \dots & & 0 & \frac{1}{2\sqrt{2}} \end{pmatrix},$$

and

$$\mathbf{M}_{j,1} = \begin{pmatrix} 1\,0\,0 & & \\ 0\,0\,0 & & \\ 0\,1\,0 & & \\ \vdots\,\vdots\,\vdots & & \vdots \\ & & 0\,1\,0 \\ & & 0\,0\,0 \\ & & 0\,0\,1 \end{pmatrix}.$$

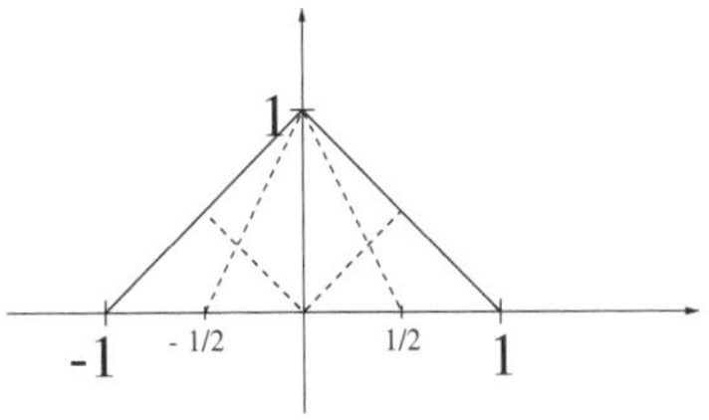

Fig. 5. Hat function

One also derives directly from Figure 5 how to express the fine scale basis functions in terms of the coarse scale and complement basis functions (the latter being simply the fine scale basis functions at the new knots):

$$
\begin{aligned}
\phi_{j+1,2k} &= \sqrt{2}\,\phi_{j,k} - \frac{1}{2}\left(\phi_{j+1,2k-1} + \phi_{j+1,2k+1}\right) \\
&= \sqrt{2}\,\phi_{j,k} - \frac{1}{2}\left(\psi_{j,k-1} + \psi_{j,k}\right), \quad k = 1,\ldots,2^j - 1, \\
\phi_{j+1,2k+1} &= \psi_{j,k}, \quad k = 0,\ldots,2^j - 1, \\
\phi_{j+1,0} &= \sqrt{2}\,\phi_{j,0} - \frac{1}{2}\psi_{j,0}, \quad \phi_{j+1,2^{j+1}} = \sqrt{2}\,\phi_{j,2^j} - \frac{1}{2}\psi_{j,2^j-1},
\end{aligned}
$$

Accordingly, the blocks of the matrix $\mathbf{G}_j$ look as follows:

$$
\mathbf{G}_{j,0} = \begin{pmatrix}
0 & \sqrt{2} & 0 & 0 & 0 & & \ldots & 0 & 0 \\
0 & 0 & 0 & \sqrt{2} & 0 & & \vdots & 0 & 0 \\
\vdots & \vdots & & & & & \vdots & & \vdots \\
0 & & & & & \sqrt{2} & 0 & 0 & 0 \\
0 & 0 & \ldots & & & 0 & 0 & \sqrt{2} & 0
\end{pmatrix}
$$

$$
\mathbf{G}_{j,1} = \begin{pmatrix}
1 & -\frac{1}{2} & 0 & 0 & \ldots & & \\
0 & -\frac{1}{2} & 1 & -\frac{1}{2} & & & \vdots \\
\vdots & & & & & & \vdots \\
\vdots & & & & & -\frac{1}{2} & 0 \\
0 & \ldots & & & & -\frac{1}{2} & 1
\end{pmatrix}.
$$

5.5 Structure of Multiscale Transformations

Recall that by (46) and (47)

$$
\Phi_j^T \mathbf{c}^j + \Psi_j^T \boldsymbol{d}^j = \Phi_{j+1}^T \left(\mathbf{M}_{j,0}\mathbf{c}^j + \mathbf{M}_{j,1}\boldsymbol{d}^j\right), \tag{50}
$$

so that $\mathbf{T}_J : \boldsymbol{d} \to \mathbf{c}$ has the form

$$\begin{array}{ccccccccc}
\mathbf{c}^0 & \stackrel{\mathbf{M}_{0,0}}{\to} & \mathbf{c}^1 & \stackrel{\mathbf{M}_{1,0}}{\to} & \mathbf{c}^2 & \stackrel{\mathbf{M}_{2,0}}{\to} & \cdots & \stackrel{\mathbf{M}_{J-1,0}}{\to} & \mathbf{c}^J \\
 & \mathbf{M}_{0,1} & & \mathbf{M}_{1,1} & & \mathbf{M}_{2,1} & & \mathbf{M}_{J-1,1} & \\
 & \nearrow & & \nearrow & & \nearrow & \cdots & \nearrow & \\
\boldsymbol{d}^0 & & \boldsymbol{d}^1 & & \boldsymbol{d}^2 & & & \boldsymbol{d}^{J-1} &
\end{array}$$

see Section 2.1. Thus the transform consists of the successive application of the matrices

$$\mathbf{T}_{J,j} := \begin{pmatrix} \mathbf{M}_j & \mathbf{0} \\ \mathbf{0} & \mathbf{I} \end{pmatrix}, \qquad \mathbf{T}_J = \mathbf{T}_{J,J-1} \cdots \mathbf{T}_{J,0}.$$

In order to determine the inverse transform $\mathbf{T}_J^{-1} : \mathbf{c} \to \boldsymbol{d}$, note that by (48),

$$\Phi_{j+1}^T \mathbf{c}^{j+1} = \Phi_j^T (\mathbf{G}_{j,0} \mathbf{c}^{j+1}) + \Psi_j^T (\mathbf{G}_{j,1} \mathbf{c}^{j+1}) = \Phi_j^T \mathbf{c}^j + \Psi_j^T \boldsymbol{d}^j,$$

which yields

$$\begin{array}{ccccccccc}
\mathbf{c}^J & \stackrel{\mathbf{G}_{J-1,0}}{\to} & \mathbf{c}^{J-1} & \stackrel{\mathbf{G}_{J-2,0}}{\to} & \mathbf{c}^{J-2} & \stackrel{\mathbf{G}_{J-3,0}}{\to} & \cdots & \stackrel{\mathbf{G}_{0,0}}{\to} & \mathbf{c}^0 \\
 & \mathbf{G}_{J-1,1} & & \mathbf{G}_{J-2,1} & & \mathbf{G}_{J-3,1} & & \mathbf{G}_{0,1} & \\
 & \searrow & & \searrow & & \searrow & \cdots & \searrow & \\
 & & \boldsymbol{d}^{J-1} & & \boldsymbol{d}^{J-2} & & & & \boldsymbol{d}^0.
\end{array}$$

5.6 Parametrization of Stable Completions

The above example of the hierarchical basis shows that often *some* stable completions are easy to construct. The corresponding complement bases may, however, not be appropriate yet, because the corresponding multiscale basis may not have a dual basis in L_2. The hierarchical basis is an example. The idea is then to *modify* some *initial* stable completion to obtain a better one. Therefore the following parametrization of all possible stable completions is of interest [23].

Theorem 3. *Given some* initial *completion* $\check{\mathbf{M}}_{j,1}$ *(and* $\check{\mathbf{G}}_j$*), then* all *other completions have the form*

$$\mathbf{M}_{j,1} = \mathbf{M}_{j,0}\mathbf{L} + \check{\mathbf{M}}_{j,1}\mathbf{K} \tag{51}$$

and

$$\mathbf{G}_{j,0} = \check{\mathbf{G}}_{j,0} - \check{\mathbf{G}}_{j,1}(\mathbf{K}^T)^{-1}\mathbf{L}^T, \quad \mathbf{G}_{j,1} = \check{\mathbf{G}}_{j,1}(\mathbf{K}^T)^{-1}. \tag{52}$$

The main point of the proof is the identity

$$\mathbf{I} = \mathbf{M}_j \mathbf{G}_j = \mathbf{M}_j \begin{pmatrix} \mathbf{I} & \mathbf{L} \\ \mathbf{0} & \mathbf{K} \end{pmatrix} \begin{pmatrix} \mathbf{I} & -\mathbf{L}\mathbf{K}^{-1} \\ \mathbf{0} & \mathbf{K}^{-1} \end{pmatrix} \mathbf{G}_j =: \check{\mathbf{M}}_j \check{\mathbf{G}}_j.$$

The special case $\mathbf{K} = \mathbf{I}$ is often referred to as *Lifting Scheme* [70]. Modifications of the above type can be used for the following purposes:

- Raising the order of *vanishing moments*: Choose $\mathbf{K} = \mathbf{I}$ and $\mathbf{L}$ such that

$$\int_\Omega \Psi_j^T P dx = \int_\Omega \Phi_{j+1}^T \mathbf{M}_{j,1} P dx$$
$$= \int_\Omega \Phi_j^T \mathbf{L} P + \check{\Psi}_j^T P dx = 0, \quad P \in \mathbb{P}_{m^*}.$$

- Construction of finite element based wavelets through coarse grid corrections [23, 40, 42, 74].
- Construction of *biorthogonal* wavelets on intervals and multivariate domains [21, 22, 30, 46, 47].

In particular, a systematic *biorthogonalization* can be described as follows. Suppose that dual pairs of single scale bases are given

$$\Phi_j^T = \Phi_{j+1}^T \mathbf{M}_{j,0}, \quad \tilde{\Phi}_j^T = \tilde{\Phi}_{j+1}^T \tilde{\mathbf{M}}_{j,0}, \quad \langle \Phi_j, \tilde{\Phi}_j \rangle = \mathbf{I}.$$

Theorem 4. *Let* $\tilde{\mathbf{M}}_{j,0}, \check{\mathbf{M}}_{j,1}, \check{\mathbf{G}}_j$ *be as above. Then* $(\mathbf{K} := \mathbf{I},\ \mathbf{L} := -\tilde{\mathbf{M}}_{j,0}^T \check{\mathbf{M}}_{j,1})$

$$\mathbf{M}_{j,1} := (I - \mathbf{M}_{j,0} \tilde{\mathbf{M}}_{j,0}^T) \check{\mathbf{M}}_{j,1}, \qquad \tilde{\mathbf{M}}_{j,1} := \check{\mathbf{G}}_{j,1}$$

are new uniformly stable completions satisfying $\mathbf{M}_j \tilde{\mathbf{M}}_j^T = \mathbf{I}$ *and*

$$\Psi_j^T := \Phi_{j+1}^T \mathbf{M}_{j,1}, \qquad \tilde{\Psi}_{j+1}^T := \tilde{\Phi}_{j+1}^T \tilde{\mathbf{M}}_{j,1}$$

form biorthogonal wavelet bases.

The criteria from Section 4 can be used to show that the constructions based on these concepts give indeed rise to Riesz bases, cf. [21, 22, 30, 46, 47]. Very useful explicit expressions for the entries of $\mathbf{L} := -\tilde{\mathbf{M}}_{j,0}^T \check{\mathbf{M}}_{j,1}$ are derived in [62]

6 Scope of Problems

We are going to describe next the scope of problems to which the above tools will be applied.

6.1 Problem Setting

Let $\mathcal{H}$ be a Hilbert space and $A(\cdot,\cdot) : \mathcal{H} \times \mathcal{H} \to I\!R$ a continuous bilinear form, i.e.,

$$|A(V,U)| \ \lesssim \ \|V\|_{\mathcal{H}} \|U\|_{\mathcal{H}}, \quad V, U \in \mathcal{H}. \tag{53}$$

We will be concerned with the variational problem: Given $F \in \mathcal{H}'$ find $U \in \mathcal{H}$ such that

$$A(V,U) = \langle V,F\rangle, \quad V \in \mathcal{H}. \tag{54}$$

We explain first what we mean when saying that (54) is *well-posed.* To this end, define the operator $\mathcal{L} : \mathcal{H} \to \mathcal{H}'$ by

$$\langle V, \mathcal{L}U\rangle = A(V,U), \quad V \in \mathcal{H}, \tag{55}$$

so that (54) is equivalent to

$$\mathcal{L}U = F. \tag{56}$$

Then (54) is called well-posed (on $\mathcal{H}$) if there exist positive finite constants $c_{\mathcal{L}}, C_{\mathcal{L}}$ such that

$$c_{\mathcal{L}}\|V\|_{\mathcal{H}} \le \|\mathcal{L}V\|_{\mathcal{H}'} \le C_{\mathcal{L}}\|V\|_{\mathcal{H}}, \quad V \in \mathcal{H}. \tag{57}$$

We will refer to (57) as *mapping property* (MP). Clearly (MP) implies the existence of a unique solution for any $F \in \mathcal{H}'$ which depends continuously on the data F (with respect to the topology of $\mathcal{H}$).

Remark 8 *It should be noted that in many cases, given $A(\cdot,\cdot)$ the first (and often most crucial) task is to* identify *a suitable Hilbert space $\mathcal{H}$ such that the mapping property (57) holds.*

All examples to be considered later will have the following format. $\mathcal{H}$ will in general be a product space

$$\mathcal{H} = H_{1,0} \times \cdots H_{m,0},$$

where the component spaces $H_{i,0} \subseteq H_i$ will be closed subspaces of Hilbert spaces H_i with norms $\|\cdot\|_{H_i}$. The H_i are typically Sobolev spaces and the $H_{i,0}$ could be determined by homogeneous boundary conditions $H_0^{t_i}(\Omega_i) \subseteq H_{i,0} \subseteq H^{t_i}(\Omega_i)$. We use capital letters to indicate that the elements of $\mathcal{H}$ have in general several components so that $V = (v_1,\ldots,v_m)^T$ and $\|V\|_{\mathcal{H}}^2 = \sum_{i=1}^m \|v_i\|_{H_i}^2$. Denoting by $\langle\cdot,\cdot\rangle_i$ a dual pairing on $H_i \times H_i'$ and setting $\langle V, W\rangle = \sum_{i=1}^m \langle v_i, w_i\rangle_i$, the dual space $\mathcal{H}'$ is endowed, as usual, with the norm

$$\|W\|_{\mathcal{H}'} = \sup_{V\in\mathcal{H}} \frac{\langle V,W\rangle}{\|V\|_{\mathcal{H}}}.$$

The bilinear form $A(\cdot,\cdot)$ will in general have the form $A(V,W) = (a_{i,l}(v_i,w_l))_{i,l=1}^m$, so that the operator $\mathcal{L}$ is matrix valued as well $\mathcal{L} = (\mathcal{L}_{i,l})_{i,l=1}^m$.

We proceed now discussing briefly several examples and typical obstructions to their numerical treatment.

6.2 Scalar 2nd Order Elliptic Boundary Value Problem

Suppose that $\Omega \subset \mathbb{R}^d$ is a bounded (Lipschitz) domain and $a(x)$ is a symmetric (bounded) matrix that is uniformly positive definite on Ω. The classical boundary value problem associated with this second order partial differential equation reads

$$-\mathrm{div}\big(a(x)\nabla u\big) + k(x)u = f \text{ on } \Omega, \quad u = 0 \text{ on } \partial\Omega. \tag{58}$$

We shall reserve lower case letters to such scalar problems. Its weak formulation has the form (54) with $m = 1$ and

$$a(v,w) := \int_\Omega a\nabla v^T \nabla w + kvw dx, \quad \mathcal{H} = H_0^1(\Omega), \quad \mathcal{H}' = H^{-1}(\Omega). \tag{59}$$

Classical finite difference or finite element discretizations turn (59) into a *finite dimensional* linear system of equations. When solving these systems, one encounters the following

Obstructions:

- The systems are sparse but in realistic cases often very large so that the use of direct solvers based on elimination techniques, is excluded. In fact, the *fill-in* caused by elimination would result in prohibitive storage and CPU demands.
- Hence one has to resort to iterative solvers whose efficiency depends on the *condition numbers* of the systems. Unfortunately, the systems are increasingly *ill-conditioned.* When the mesh size h decreases, $\mathrm{cond}_2(a(\Phi_h, \Phi_h)) \sim h^{-2}$, where $a(\Phi_h, \Phi_h)$ denotes the stiffness matrix with respect to an L_2-stable single scale basis, such as a standard nodal finite element basis.

6.3 Global Operators – Boundary Integral Equations

Let Ω^- be again a bounded domain in $\mathbb{R}^d$ $(d \in \{2,3\})$ and consider the following special case of (58)

$$-\Delta w = 0, \text{ on } \Omega, \quad (\Omega = \Omega^- \text{ or } \Omega^+ := \mathbb{R}^3 \setminus \Omega^-), \tag{60}$$

subject to the boundary conditions

$$w = f \quad \text{on } \Gamma := \partial\Omega^- \quad (w(x) \to 0, \ |x| \to \infty \text{ when } \Omega = \Omega^+). \tag{61}$$

Of course, the unbounded domain Ω^+ poses an additional difficulty in the case of such an *exterior* boundary value problem. A well-known strategy is to transform (60), (61) into a *boundary integral equation* that lives only on the manifold Γ. There are several ways to do that. They all involve the fundamental solution of the Laplace operator $\mathcal{E}(x,y) = 1/4\pi|x-y|$ which gives rise to the *single layer potential* operator

$$(\mathcal{L}u)(x) = (\mathcal{V}u)(x) := \int_\Gamma \mathcal{E}(x,y)u(y)d\Gamma_y, \quad x \in \Gamma. \tag{62}$$

One can then show that the solution u of the first kind integral equation

$$\mathcal{V}u = f \quad \text{on} \quad \Gamma \tag{63}$$

provides the solution w of (60) through the representation formula

$$w(x) = \int_\Gamma \mathcal{E}(x,y)u(y)d\Gamma_y, \quad x \in \Omega.$$

Here one has (see e.g.[64])

$$a(v,u) = \langle v, \mathcal{V}u \rangle_\Gamma, \quad \mathcal{H} = H^{-1/2}(\Gamma), \ \mathcal{H}' = H^{1/2}(\Gamma).$$

An alternative way uses the *double layer potential*

$$(\mathcal{K}v)(x) := \int_\Gamma \frac{\partial}{\partial n_y}\mathcal{E}(x,y)v(y)d\Gamma_y = \int_\Gamma \frac{1}{4\pi}\frac{\nu_y^T(x-y)}{|x-y|^3}v(y)\,d\Gamma_y, \quad x \in \Gamma, \tag{64}$$

where ν is the outward normal to Γ. Now the solution of the second kind integral equation

$$\mathcal{L}u := (\frac{1}{2} \pm \mathcal{K})u = f \quad (\Omega = \Omega^\pm) \tag{65}$$

gives the solution to (60) through

$$w(x) = \int_\Gamma \mathcal{K}(x,y)u(y)d\Gamma_y.$$

In this case the bilinear form and the corresponding energy space are as follows

$$a(v,w) = \langle v, (\frac{1}{2} \pm \mathcal{K})w \rangle_\Gamma, \quad \mathcal{H} = L_2(\Gamma) = H_2 = \mathcal{H}'.$$

The so called hypersingular operator offers yet another alternative in which case $\mathcal{H}$ turns out to be $H^{1/2}(\Gamma)$. According to the shifts caused by these operators in the Sobolev scale the single layer potential, double layer potential and hypersingular operator have order $-1, 0, 1$, respectively.

In a similar way Neumann boundary conditions can be treated. Moreover, other classical elliptic systems can be treated similarly where the above operators serve as core ingredients.

The obvious advantage of the above approach is the reduction of the spatial dimension and that one has to discretize in all cases only bounded domains. On the other hand, one faces the following

Obstructions:

- Discretizations lead in general to *densely populated matrices*. This severely limits the number of degrees of freedom when using direct solvers. But even iterative techniques are problematic, due to the cost of the matrix/vector multiplication.
- Whenever the order of the operator is different from zero, the problem of *growing condition numbers* arises (e.g. $\mathcal{L} = \mathcal{V}$), see Section 10.

6.4 Saddle Point Problems

All the above examples involve scalar coercive bilinear forms. An important class of problems which are no longer coercive are *saddle point problems*. A detailed treatment of this type of problems can be found in [19, 59]. Suppose X, M are Hilbert spaces and that $a(\cdot,\cdot)$, $b(\cdot,\cdot)$ are bilinear forms on $X \times X$, respectively $X \times M$ which are continuous

$$|a(v,w)| \lesssim \|v\|_X \|w\|_X, \quad |b(q,v)| \lesssim \|v\|_X \|q\|_M. \tag{66}$$

Given $f \in X'$, $g \in M'$, find $U = (u,p) \in X \times M =: \mathcal{H}$ such that one has for all $V = (v,q) \in \mathcal{H}$

$$A(U,V) = \begin{cases} a(u,v) + b(p,v) = \langle f, v\rangle, \\ b(q,u) \qquad\qquad\; = \langle q, g\rangle. \end{cases} \tag{67}$$

Note that when $a(\cdot,\cdot)$ is positive definite symmetric, the solution component u minimizes the quadratic functional $J(w) := \frac{1}{2}a(w,w) - \langle f, w\rangle$ subject to the constraint $b(u,q) = \langle q, g\rangle$, for all $q \in M$, i.e.,

$$\inf_{v\in X} \sup_{q\in M} \left(\frac{1}{2} a(v,v) + b(v,q) - \langle f, v\rangle - \langle g, q\rangle \right).$$

This accounts for the term saddle point problem (even under more general assumptions on $a(\cdot,\cdot)$).

In order to write (67) as an operator equation, define the operators A, B by

$$a(v,w) =: \langle v, Aw\rangle, \quad v \in X, \quad b(v,p) =: \langle Bv, q\rangle, \quad q \in M,$$

so that (67) becomes

$$\mathcal{L}U := \begin{pmatrix} A & B' \\ B & 0 \end{pmatrix} \begin{pmatrix} u \\ p \end{pmatrix} = \begin{pmatrix} f \\ g \end{pmatrix} =: F. \tag{68}$$

As for the mapping property (MP) (57), a simple (sufficient) condition reads as follows [19, 59]. If $a(\cdot,\cdot)$ is *elliptic* on

$$\ker B := \{v \in X : b(v,q) = 0, \ \forall\, q \in M\},$$

i.e.,

$$a(v,v) \sim \|v\|_X^2, \quad v \in \ker B, \tag{69}$$

and if $b(\cdot,\cdot)$ satisfies the *inf-sup condition*

$$\inf_{q\in M} \sup_{v\in X} \frac{b(v,q)}{\|v\|_X \|q\|_M} > \beta \tag{70}$$

for some positive β, then (66) is well-posed, i.e., $\mathcal{L}$ defined by (68) satisfies

$$c_{\mathcal{L}} \left(\|v\|_X^2 + \|q\|_M^2\right)^{1/2} \le \|\mathcal{L}\binom{v}{q}\|_{X'\times M'} \le C_{\mathcal{L}} \left(\|v\|_X^2 + \|q\|_M^2\right)^{1/2}. \tag{71}$$

Condition (70) means that B is surjective (and thus has closed range). Condition (69) is actually too strong. It can be replaced by requiring bijectivity of A on $\ker B$, see [19], which will be used in some of the examples below. Before turning to these examples we summarize some principal

Obstructions:

- As in previous examples discretizations lead usually to large linear systems that become more and more ill-conditioned when the resolution of the discretization increases.
- An additional difficulty is caused by the fact that the form (67) is *indefinite*, so that more care has to be taken when devising an iterative scheme.
- An important point is that the well-posedness of the infinite dimensional problem (71) is *not* automatically inherited by a finite dimensional Galerkin discretization. In fact, the trial spaces in X and M have to be compatible in the sense that they satisfy the inf-sup condition (70) *uniformly* with respect to the resolution of the chosen discretizations. This is called the *Ladyšhenskaja-Babuška-Brezzi-condition* (LBB) and may, depending on the problem, be a delicate task.

  We discuss next some special cases.

Second Order Problem - Fictitious Domains $\Omega \subset \Box$:

Instead of incorporating boundary conditions for the second order problem (58) in the finite dimensional trial spaces, one can treat them as constraints for a variational problem, that is formulated over some possibly larger but simpler domain $\Box$, e.g. a cube. This is of interest when the boundary varies or when boundary values are used as control variables. Appending these constraints with the aid of Lagrange multipliers, leads to the following saddle point problem, cf. [16, 43, 60].

Find $U = (u,p) \in \mathcal{H} := H^1(\Box) \times H^{-1/2}(\Gamma)$, $\Gamma := \partial\Omega$, such that

$$\begin{aligned} \langle \nabla v, \mathbf{a}\nabla u\rangle + \langle v,p\rangle_\Gamma &= \langle v,f\rangle \qquad &&\text{for all } v \in H^1(\Box), \\ \langle q,u\rangle_\Gamma &= \langle g,q\rangle &&\text{for all } q \in H^{-1/2}(\Gamma). \end{aligned} \tag{72}$$

Remark 9 *The problem (72) is well posed, i.e., (71) holds.*

Proof: Clearly, (69) is satisfied. Moreover, the inf-sup condition (70) is a consequence of the *'inverse' Trace Theorem* that states that

$$\inf_{v \in H^1(\Omega), v|_{\partial\Omega}=g} \|v\|_{H^1(\Omega)} \lesssim \|g\|_{H^{1/2}(\partial\Omega)},$$

[19, 59]. In fact, in the present case we have $b(v,p) = \langle v,p\rangle_\Gamma$. Given $p \in H^{-1/2}(\Gamma)$, choose $g \in H^{1/2}(\Gamma)$ such that

$$\|p\|_{H^{-1/2}(\Gamma)} \le 2\langle g,p\rangle_\Gamma / \|g\|_{H^{1/2}(\Gamma)} \lesssim \langle v,p\rangle_\Gamma / \|v\|_{H^1(\Omega)} \lesssim \langle v,p\rangle_\Gamma / \|v\|_{H^1(\Box)}$$

for some $v \in H^{(}\Box)$. Thus $b(v,p) \gtrsim \|v\|_{H^1(\Box)} \|p\|_{H^{-1/2}(\Gamma)}$ which confirms the claim. □

First Order Systems

One is often more interested in derivatives of the solution to the boundary value problem (58). Introducing the *fluxes* $\boldsymbol{\theta} := -\mathbf{a}\nabla u$, (58) can be written as a system of first order equations whose weak formulations reads

$$\begin{aligned} \langle \boldsymbol{\theta}, \boldsymbol{\eta}\rangle + \langle \boldsymbol{\eta}, \mathbf{a}\nabla u\rangle &= 0, \qquad \forall\, \boldsymbol{\eta} \in \mathbf{L}_2(\Omega), \\ -\langle \boldsymbol{\theta}, \nabla v\rangle + \langle ku, v\rangle \quad &= \langle f, v\rangle, \ \forall v \in H^1_{0,\Gamma_D}(\Omega). \end{aligned} \tag{73}$$

One now looks for a solution

$$U = (\boldsymbol{\theta}, u) \ \in \ \mathcal{H} := \mathbf{L}_2(\Omega) \times H^1_{0,\Gamma_D}(\Omega), \tag{74}$$

where $H^1_{0,\Gamma_D}(\Omega)$ is the closure in $H^1(\Omega)$ of all smooth functions whose support does not intersect Γ_D. For a detailed discussion in the finite element context see e.g. [17, 18]. It turns out that in this case the Galerkin discretization inherits the stability from the original second order problem.

The Stokes System

The simplest model for viscous incompressible fluid flow is the Stokes system

$$\begin{aligned} -\nu\Delta\mathbf{u} + \nabla p = \mathbf{f} \qquad &\text{in } \ \Omega, \\ \operatorname{div} \mathbf{u} = 0 \qquad &\text{in } \ \Omega, \\ \mathbf{u}|_\Gamma = \mathbf{0}, \qquad & \end{aligned} \tag{75}$$

where $\mathbf{u}$ and p are the velocity, respectively pressure, see [19, 59]. The relevant function spaces are

$$X = \mathbf{H}^1_0(\Omega) := (H^1_0(\Omega))^d, \quad M = L_{2,0}(\Omega) := \{q \in L_2(\Omega) : \int_\Omega q = 0\}. \tag{76}$$

In fact, one can show that the range of the divergence operator is $L_{2,0}(\Omega)$. The weak formulation of (75) is

$$\begin{aligned} \nu\langle\nabla\mathbf{v},\nabla\mathbf{u}\rangle_{\mathbf{L}_2(\Omega)} + \langle\,\mathrm{div}\;\mathbf{v},p\rangle_{L_2(\Omega)} &= \langle\mathbf{f},\mathbf{v}\rangle,\ \mathbf{v}\in\mathbf{H}_0^1(\Omega)\\ \langle\,\mathrm{div}\;\mathbf{u},q\rangle_{L_2(\Omega)} &= \mathbf{0},\qquad q\in L_{2,0}(\Omega), \end{aligned} \tag{77}$$

i.e., one seeks a solution $U=(\mathbf{u},p)$ in the energy space

$$\mathcal{H} = X\times M = \mathbf{H}_0^1(\Omega)\times L_{2,0}(\Omega),$$

for which the mapping property (57) can be shown to hold [19, 59].

Stokes System - Fictitious Domain Formulation

An example with $m=3$ components is obtained by weakly enforcing inhomogeneous boundary conditions for the Stokes system (see [61])

$$\begin{aligned} -\nu\Delta\mathbf{u}+\nabla p &= \mathbf{f} \qquad \text{in } \Omega,\\ \mathrm{div}\;\mathbf{u} &= 0 \qquad \text{in } \Omega,\\ \mathbf{u}|_\Gamma &= \mathbf{g},\\ \int_\Omega p\,dx &= 0,\quad \Big(\int_\Gamma \mathbf{g}\cdot\mathbf{n}\,ds = 0\Big), \end{aligned}$$

whose weak formulation is

$$\begin{aligned} \nu\langle\nabla\mathbf{v},\nabla\mathbf{u}\rangle_{\mathbf{L}_2(\Omega)} + \langle\mathbf{v},\boldsymbol{\lambda}\rangle_{\mathbf{L}_2(\Gamma)} + \langle\,\mathrm{div}\;\mathbf{v},p\rangle_{L_2(\Omega)} &= \langle\mathbf{f},\mathbf{v}\rangle\ \forall\ \mathbf{v}\in\mathbf{H}^1(\Omega),\\ \langle\mathbf{u},\boldsymbol{\mu}\rangle_{\mathbf{L}_2(\Gamma)} &= \langle\mathbf{g},\boldsymbol{\mu}\rangle\ \forall\ \boldsymbol{\mu}\in\mathbf{H}^{-1/2}(\Gamma),\\ \langle\,\mathrm{div}\;\mathbf{u},q\rangle_{L_2(\Omega)} &= \mathbf{0}\qquad\forall\ q\in L_{2,0}(\Omega). \end{aligned}$$

The unknowns are now velocity, pressure and an additional Lagrange multiplier for the boundary conditions $U=(\mathbf{u},\boldsymbol{\lambda},p)$. An appropriate energy space is

$$\mathcal{H} := \mathbf{H}^1(\Omega)\times\mathbf{H}^{-1/2}(\Gamma)\times L_{2,0}(\Omega).$$

Well-posedness can either be argued by taking $b(\mathbf{v},(\boldsymbol{\lambda},p)) = \langle\mathbf{v},\boldsymbol{\lambda}\rangle_{\mathbf{L}_2(\Gamma)} + \langle\,\mathrm{div}\;\mathbf{v},p\rangle_{L_2(\Omega)}$ and observing that (69) holds, or by taking $b(\mathbf{v},\boldsymbol{\lambda})=\langle\mathbf{v},\boldsymbol{\lambda}\rangle_{\mathbf{L}_2(\Gamma)}$ and using the fact that the Stokes operator is bijective on the kernel of this constraint operator.

First Order Stokes System

The same type of argument can be used to show that the following first order formulation of the Stokes system leads to a well-posed variational problem (with $m=4$), [20, 44]. The fluxes $\underline{\boldsymbol{\theta}}$ are now matrix valued indicated by the underscore:

$$\begin{aligned} \underline{\boldsymbol{\theta}} + \nabla \mathbf{u} &= \underline{\mathbf{0}} \text{ in } \Omega, \\ -\nu(\operatorname{div} \underline{\boldsymbol{\theta}})^T + \nabla p &= \mathbf{f} \text{ in } \Omega, \\ \operatorname{div} \mathbf{u} &= 0 \text{ in } \Omega, \\ \mathbf{u} &= \mathbf{g} \text{ on } \Gamma. \end{aligned}$$

The unknowns are now $U = (\underline{\boldsymbol{\theta}}, \mathbf{u}, p, \boldsymbol{\lambda}) \in \mathcal{H}$ where the energy space

$$\mathcal{H} := \underline{\mathbf{L}}_2(\Omega) \times \mathbf{H}^1(\Omega) \times L_{2,0}(\Omega) \times \mathbf{H}^{-1/2}(\Gamma),$$

can be shown to render the following weak formulation

$$\begin{array}{llll} \langle \underline{\boldsymbol{\theta}}, \underline{\boldsymbol{\eta}} \rangle & +\langle \underline{\boldsymbol{\eta}}, \nabla \mathbf{u} \rangle & = 0, & \underline{\boldsymbol{\eta}} \in \underline{\mathbf{L}}_2(\Omega), \\ \nu \langle \underline{\boldsymbol{\theta}}, \nabla \mathbf{v} \rangle & \quad -\langle p, \operatorname{div} \mathbf{v} \rangle - \langle \boldsymbol{\lambda}, \mathbf{v} \rangle_\Gamma & = \langle \mathbf{f}, \mathbf{v} \rangle, & \mathbf{v} \in \mathbf{H}^1(\Omega), \\ & \langle \operatorname{div} \mathbf{u}, q \rangle & = 0, & q \in L_{2,0}(\Omega), \\ & \langle \boldsymbol{\mu}, \mathbf{u} \rangle_\Gamma & = \langle \boldsymbol{\mu}, \mathbf{g} \rangle_\Gamma, & \boldsymbol{\mu} \in \mathbf{H}^{-1/2}(\Gamma), \end{array}$$

satisfy the mapping property (57).

Transmission Problem

The following example is interesting because it involves both local and global operators [31]

$$\begin{aligned} -\nabla \cdot (\mathbf{a} \nabla u) &= f \qquad \text{in } \Omega_0, \\ -\Delta u &= 0 \qquad \text{in } \Omega_1, \\ u|_{\Gamma_0} &= 0 \\ \mathcal{H} &:= H^1_{0,\Gamma_D}(\Omega_0) \times H^{-1/2}(\Gamma_1). \end{aligned}$$

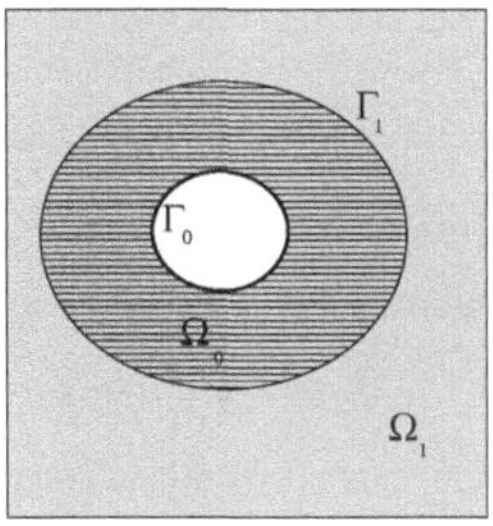

Both boundary value problems are coupled by the *interface conditions:*

$$u^- = u^+, \quad (\partial_{\mathbf{n}})u^- = (\partial_{\mathbf{n}})u^+.$$

A well-posed weak formulation of this problem with respect to the above $\mathcal{H}$ is

$$\langle \mathbf{a} \nabla u, \nabla v \rangle_{\Omega_0} + \langle \mathcal{W} u - (\tfrac{1}{2}\mathcal{I} - \mathcal{K}')\sigma, v \rangle_{\Gamma_1} = \langle f, v \rangle_{\Omega_0}, \qquad v \in H^1_{0,\Gamma_D}(\Omega_0),$$

$$\langle (\tfrac{1}{2}\mathcal{I} - \mathcal{K})u, \delta \rangle_{\Gamma_1} + \langle \mathcal{V}\sigma, \delta \rangle_{\Gamma_1} = 0, \qquad \delta \in H^{-1/2}(\Gamma_1),$$

where $\mathcal{W}$ denotes here the hypersingular operator, see [31, 44].

Note that in all these examples, as an additional obstruction, the occurrence and evaluation of *difficult norms* $\|\cdot\|_{H^{1/2}(\Gamma)}$, $\|\cdot\|_{H^{-1/2}(\Gamma)}$, $\|\cdot\|_{H^{-1}(\Omega)}$ arises.

7 An Equivalent ℓ_2-Problem

We shall now describe how the wavelet concepts from Section 3 apply to the above type of problems.

An important distinction from conventional approaches lies in the fact, that for suitable choices of wavelet bases the original problem (54) can be transformed into an equivalent one defined in ℓ_2. Moreover, it will turn out to be well-posed in Euclidean metric. Here 'suitable' means that the component spaces of $\mathcal{H}$ admit a *wavelet characterization* in the sense of (16). Specifically, we will assume that for each $H_{i,0}$ one has suitable bases Ψ^i and scaling matrices $\mathbf{D}_i$ such that

$$c_\Psi \|\mathbf{v}\|_{\ell_2(\mathcal{J}_i)} \leq \|\mathbf{v}^T \mathbf{D}_i^{-1} \Psi^i\|_{H_i} \leq C_\Psi \|\mathbf{v}\|_{\ell_2(\mathcal{J}_i)}, \quad \mathbf{v} \in \ell_2,\ i = 1, \dots, m. \tag{78}$$

It will be convenient to adopt the following notational conventions.

$$\mathcal{J} := \mathcal{J}_1 \times \cdots \times \mathcal{J}_m, \quad \mathbf{D} := \operatorname{diag}\left(\mathbf{D}_1, \dots, \mathbf{D}_m\right), \quad \mathbf{V} = (\mathbf{v}^1, \dots, \mathbf{v}^m)^T.$$

Thus, catenating the quantities corresponding to the component spaces, we will employ the following compact notation

$$\mathbf{V}^T \mathbf{D}^{-1} \boldsymbol{\Psi} := ((\mathbf{v}^1)^T \mathbf{D}_1^{-1} \Psi^1, \dots, (\mathbf{v}^m)^T \mathbf{D}_m^{-1} \Psi^m)^T.$$

In these terms the *wavelet characterization of the energy space* can be expressed as

$$c_\Psi \|\mathbf{V}\|_{\ell_2} \leq \|\mathbf{V}^T \mathbf{D}^{-1} \boldsymbol{\Psi}\|_{\mathcal{H}} \leq C_\Psi \|\mathbf{V}\|_{\ell_2}. \tag{79}$$

Recalling Section 6.1, the (scaled) wavelet representation of the operators $\mathcal{L}_{i,l}$, defined by (55), is given by

$$\boldsymbol{A}_{i,l} := \mathbf{D}_i^{-1} a_{i,l}(\Psi^i, \Psi^l) \mathbf{D}_l^{-1}, \quad i, l = 1, \dots, m.$$

The *scaled standard representation* of $\mathcal{L}$, defined by (55), and the dual wavelet representation of the right hand side data are given by

$$\mathbf{L} := \left(\boldsymbol{A}_{i,l}\right)_{i,l=1}^m = \mathbf{D}^{-1} \langle \boldsymbol{\Psi}, \mathcal{L} \boldsymbol{\Psi} \rangle \mathbf{D}^{-1}, \quad \mathbf{F} := \mathbf{D}^{-1} \langle \boldsymbol{\Psi}, F \rangle.$$

We shall make essential use of the following fact.

Theorem 5. *Suppose that $U = \mathbf{U}^T \mathbf{D}^{-1} \Psi$ is the scaled wavelet representation of the solution to (54). Then one has*

$$\mathcal{L}U = F \iff \mathbf{L}\mathbf{U} = \mathbf{F}, \tag{80}$$

and there exist positive constants c_L, C_L such that

$$c_L \|\mathbf{V}\|_{\ell_2} \leq \|\mathbf{L}\mathbf{V}\|_{\ell_2} \leq C_L \|\mathbf{V}\|_{\ell_2}, \quad \mathbf{V} \in \ell_2. \tag{81}$$

In fact, lower respectively upper estimates for these constants are $c_L \geq c_\Psi^2 c_{\mathcal{L}}$, $C_L \leq C_\Psi^2 C_{\mathcal{L}}$.

Proof: The proof follows from (79) and (57). In fact, let $V = \mathbf{V}^T\mathbf{D}^{-1}\boldsymbol{\Psi}$. Then

$$\begin{aligned}\|\mathbf{V}\|_{\ell_2} &\leq c_\Psi^{-1}\|V\|_{\mathcal{H}} \overset{\text{(MP)}}{\leq} c_\Psi^{-1}c_{\mathcal{L}}^{-1}\|\mathcal{L}V\|_{\mathcal{H}'} \overset{(19)}{\leq} c_\Psi^{-2}c_{\mathcal{L}}^{-1}\|\mathbf{D}^{-1}\langle\boldsymbol{\Psi},\mathcal{L}V\rangle\|_{\ell_2}\\ &= c_\Psi^{-2}c_{\mathcal{L}}^{-1}\|\mathbf{D}^{-1}\langle\boldsymbol{\Psi},\mathcal{L}\boldsymbol{\Psi}\rangle\mathbf{D}^{-1}\mathbf{V}\|_{\ell_2} = c_\Psi^{-2}c_{\mathcal{L}}^{-1}\|\mathbf{L}\mathbf{V}\|_{\ell_2}.\end{aligned}$$

The converse estimate works analogously in reverse order. □

7.1 Connection with Preconditioning

The above result can be related to preconditioning as follows. Denote by Λ some finite subset of $\mathcal{J}$. Let $\boldsymbol{\Psi}_\Lambda := \{\psi_\lambda : \lambda \in \Lambda \subset \mathcal{J}\}$ so that $S_\Lambda := \operatorname{span} \boldsymbol{\Psi}_\Lambda$ is a finite dimensional subspace of $\mathcal{H}$. Let $\mathbf{D}_\Lambda$ denote the finite principal section of $\mathbf{D}$ determined by Λ. The *stiffness matrix* of $\mathcal{L}$ with respect to $\mathbf{D}_\Lambda^{-1}\boldsymbol{\Psi}_\lambda$ is then given by $\mathbf{L}_\Lambda := \mathbf{D}_\Lambda^{-1}A(\boldsymbol{\Psi}_\Lambda, \boldsymbol{\Psi}_\Lambda)\mathbf{D}_\Lambda^{-1}$. The following observation is obvious.

Remark 10 *If $A(\cdot,\cdot)$ in (54) is $\mathcal{H}$-elliptic, then $\mathbf{L}$ is symmetric positive definite (s.p.d.) which, in turn, implies that*

$$\operatorname{cond}_2(\mathbf{L}_\Lambda) \leq \frac{C_\Psi^2 C_{\mathcal{L}}}{c_\Psi^2 c_{\mathcal{L}}}. \tag{82}$$

Ellipticity implies stability of Galerkin discretizations which explains why the infinite dimensional result (81) implies (82). The following observation stresses that it is indeed the stability of the Galerkin discretization that together with (81) implies (82).

Remark 11 *If the Galerkin discretizations with respect to the trial spaces S_Λ are stable, i.e., $\|\mathbf{L}_\Lambda^{-1}\|_{\ell_2\to\ell_2} = \mathcal{O}(1)$, $\#\Lambda \to \infty$, then one also has*

$$\|\mathbf{V}_\Lambda\|_{\ell_2} \sim \|\mathbf{L}_\Lambda\mathbf{V}_\Lambda\|_{\ell_2}, \quad \mathbf{V}_\Lambda \in \mathbb{R}^{\#\Lambda}. \tag{83}$$

Hence the mapping property (MP) of $\mathcal{L}$ and the norm equivalences (NE) imply uniformly bounded condition numbers

$$\operatorname{cond}_2(\mathbf{L}_\Lambda) = \mathcal{O}(1), \quad \#\Lambda \to \infty,$$

whenever the Galerkin discretizations with respect to S_Λ are (uniformly) stable.

Proof: By stability of the Galerkin discretizations, we only have to verify the uniform boundedness of the $\mathbf{L}_\Lambda$. Similarly as before we obtain

$$\begin{aligned}\|\mathbf{L}_\Lambda\mathbf{V}_\Lambda\|_{\ell_2} &= \|\mathbf{D}_\Lambda^{-1}A(\boldsymbol{\Psi}_\Lambda,\boldsymbol{\Psi}_\Lambda)\mathbf{D}_\Lambda^{-1}\mathbf{V}_\Lambda\|_{\ell_2} = \|\mathbf{D}_\Lambda^{-1}\langle\boldsymbol{\Psi}_\Lambda,\mathcal{L}V_\Lambda\rangle\|_{\ell_2}\\ &\leq \|\mathbf{D}^{-1}\langle\boldsymbol{\Psi},\mathcal{L}V_\Lambda\rangle\|_{\ell_2} \overset{(19)}{\leq} c_\Psi^{-1}\|\mathcal{L}V_\Lambda\|_{\mathcal{H}'}\\ &\leq c_\Psi^{-1}C_{\mathcal{L}}\|V_\Lambda\|_{\mathcal{H}} \overset{(79)}{\leq} c_\Psi^{-1}C_{\mathcal{L}}C_\Psi\|\mathbf{V}_\Lambda\|_{\ell_2},\end{aligned}$$

which proves the assertion. □

Recall that Galerkin stability for indefinite problems is in general not guaranteed for any choice of trial spaces.

7.2 There is always a Positive Definite Formulation – Least Squares

Once a well-posed problem on ℓ_2 is given, squaring will yield a symmetric positive definite formulation.

Theorem 6. *Adhering to the previous notation, let* $\mathbf{M} := \mathbf{L}^T\mathbf{L}, \quad \mathbf{G} := \mathbf{L}^T\mathbf{F}$. *Then with* $U = \mathbf{U}^T\mathbf{D}^{-1}\boldsymbol{\Psi}$ *one has*

$$\mathcal{L}U = F \quad \Longleftrightarrow \quad \mathbf{MU} = \mathbf{F},$$

and

$$c_L^{-2}\|\mathbf{V}\|_{\ell_2} \leq \|\mathbf{MV}\|_{\ell_2} \leq C_L^2\|\mathbf{V}\|_{\ell_2}.$$

Note that $\mathcal{L}U = F$ if and only if U minimizes

$$\|\mathcal{L}V - F\|_{\mathcal{H}'} \sim \|\mathbf{MV} - \mathbf{G}\|_{\ell_2}.$$

The left expression corresponds to the *natural norm least squares* formulation of (54), see [44]. Such least squares formulations have been studied extensively in the finite element context, [17, 18]. A major obstruction then stems from the fact that the dual norm $\|\cdot\|_{\mathcal{H}'}$ is often numerically hard to handle, see the examples in Section 6 involving broken trace norms or the H^{-1}-norm. Once suitable wavelet bases are available these norms become just weighted ℓ_2-norms in the transformed domain. Nevertheless, finite dimensional Galerkin approximations would still require approximating corresponding infinite sums which again raises the issue of stability. This is analyzed in [44]. However, we will see below that this problem disappears, when employing adaptive solution techniques.

8 Adaptive Wavelet Schemes

8.1 Introductory Comments

A natural way of applying the tools developed so far is to choose a finite subset $\boldsymbol{\Psi}_\Lambda$ of wavelets and solve for $U_\Lambda \in S_\Lambda$, satisfying

$$A(V, U_\Lambda) = \langle V, F\rangle, \quad V \in S_\Lambda. \tag{84}$$

The perhaps best understood case corresponds to choosing $\Lambda = (J)$, which is the set of *all* indices in $\mathcal{J}$ up to level J. For instance, when using piecewise polynomial wavelets, $\boldsymbol{\Psi}_{(J)}$ spans a finite dimensional space of piecewise polynomials on a mesh with mesh size $h \sim 2^{-J}$. For this type of trial spaces classical error estimates are available. For instance, in a scalar case ($m = 1$) when $\mathcal{H} = H^t$ one has

$$\|u - u_{(J)}\|_{H^t} \lesssim 2^{-J(s-t)}\|u\|_{H^s}, \quad J \to \infty, \tag{85}$$

provided that the solution u belongs to H^s. One can then choose J so as to meet some target accuracy ϵ. However, when the highest Sobolev regularity s of u exceeds t only by a little, one has to choose J correspondingly large, which amounts to a possibly very large number of required degrees of freedom $N \sim 2^{dJ}$. Of course, it is then very important to have efficient ways of solving the corresponding large systems of equations. This issue has been studied extensively in the literature. Suitable preconditioning techniques, based e.g. on the observations in Section 7.1, combined with *nested iteration*, (i.e., solving on successively finer discretization levels while using the approximation from the current level as initial guess on the next level,) allows one to solve the linear systems within discretization error accuracy at a computational expense that stays proportional to the number N of degrees of freedom. Thus, under the above circumstances, the amount of work needed to achieve accuracy ϵ remains *proportional* to the order of $\epsilon^{-d/(s-t)}$ degrees of freedom, which is the larger the closer s is to t. In other words, relative to the target accuracy, the computational work related to preassigned uniform discretizations is the larger the lower the *regularity*, when measured in the *same* metric as the error.

Our point of view here will be different. We wish to be as economical as possible with the number of degrees of freedom so as to still achieve some desired target accuracy. Thus instead of choosing a discretization in an a-priori manner, we wish to *adapt* the discretization to the particular case at hand. Given any target accuracy ϵ, the goal then would be to identify *ideally* a possibly small set $\Lambda(\epsilon) \subset \mathcal{J}$ and a vector $\mathbf{U}(\epsilon)$ with support $\Lambda(\epsilon)$ such that

$$\|U - \mathbf{U}(\epsilon)^T \mathbf{D}_{\Lambda(\epsilon)}^{-1} \boldsymbol{\Psi}_{\Lambda(\epsilon)}\|_{\mathcal{H}} \leq \epsilon. \tag{86}$$

The following two questions immediately come to mind.

(I) What can be said about the relation between $\#\Lambda(\epsilon)$ as a measure for the *minimum computational complexity* and the accuracy ϵ, when one has *complete* knowledge about U. We will refer to this as the *optimal work/accuracy balance* corresponding to the *best N-term approximation*

$$\sigma_{N,\mathcal{H}}(U) := \inf_{\#\Lambda \leq N, \mathbf{V}, \operatorname{supp} \mathbf{V} = \Lambda} \|U - \mathbf{V}^T \mathbf{D}_{\Lambda}^{-1} \boldsymbol{\Psi}_{\Lambda}\|_{\mathcal{H}}, \tag{87}$$

that minimizes the error in $\mathcal{H}$ for any given number N of degrees of freedoms. This is a question arising in *nonlinear approximation*, see e.g. [51]. Precise answers are known for various spaces $\mathcal{H}$. Roughly speaking the asymptotic behavior of the error of *best N-term approximation* in the above sense is governed by certain regularity scales.

(II) Best N-term approximation is in the present context only an *ideal bench mark* since U is not known. The question then is: Can one devise algorithms that are able to track during the solution process approximately the significant coefficients of U, so that the resulting work/accuracy balance is, up to a uniform constant, asymptotically the same as that of best N-term approximation?

A scheme that tracks in the above sense the significant coefficients of U, based at every stage on knowledge acquired during prior calculations, is called *adaptive.*

8.2 Adaptivity from Several Perspectives

Adaptive resolution concepts have been the subject of numerous studies from several different perspectives. In the theory of *Information Based Complexity* the performance of adaptivity versus nonadaptive strategies has been studied for a fairly general framework [65, 72]. For the computational model assumed in this context the results are not in favor of adaptive techniques.

On the other hand, adaptive *local mesh refinements*, based on *a-posteriori error estimators/indicators* in the finite element context, indicate a very promising potential of such techniques [56, 4, 5, 15, 57]. However, there seem to be no rigorous work/accuracy balance estimates that support the gained experiences.

Likewise there have been numerous investigations of adaptive wavelet schemes that are also not backed by a complexity analysis, see e.g. [2, 12, 14, 30, 35, 37].

More recently, significant progress was made by multiresolution compression techniques for hyperbolic conservation laws initiated by the work of A. Harten and continued by R. Abgral, F. Arrandiga, A. Cohen, W. Dahmen, R. DeVore, R. Donat, R. Sjorgreen, T. Sonar and others. This line was developed into a fully adaptive technique by A. Cohen, O. Kaber, S. Müller, M. Postel. The approach is based on a perturbation analysis applied to the compressed array of cell averages.

Another direction of recent developments concerns the scope of problems described in Section 6. In this context first asymptotically optimal work/accuracy balances have been established in [26, 27, 36] and will be addressed in more detail below.

8.3 The Basic Paradigm

The *classical approach* to the numerical treatment of operator equations may be summarized as follows. Starting with a variational formulation, the choice of finite dimensional trial and test spaces determines a dicretization of the continuous problem which eventually leads to a *finite dimensional problem.* The issue then is to develop efficient solvers for such problems.

As indicated in Section 6, typical obstructions are then the size of the systems, ill-conditioning, as well as compatibility constraints like the *LBB condition.*

In connection with wavelet based schemes a different paradigm suggests itself, reflected by the following steps, [27].

(I) One starts again with a variational formulation but puts first most emphasis on the *mapping property* (MP) (cf. (57)), as exemplified in Section 6.4.
(II) Instead of turning to a finite dimensional approximation, the continuous problem is transformed into an *equivalent* ∞ - *dimensional* ℓ_2 problem which is well-conditioned, see Section 7.
(III) One then tries to devise a *convergent iteration* for the ∞ - *dimensional* ℓ_2-problem.
(IV) This iteration is, of course, only conceptual. Its numerical realization relies on the *adaptive application* of the involved operators.

In this framework adaptivity enters only through ways of applying infinite dimensional operators within some accuracy tolerance. Moreover, all algorithmic steps take place in ℓ_2. The choice of the wavelet basis fully encodes the original problem, in particular, the underlying geometry. Of course, the realization of appropriate bases by itself may be a difficult problem depending on the case at hand. We will assume throughout the remainder of the paper that bases with the features described in Sections 3, 7 are available.

Thus steps (I) and (II) above have been already discussed before in Sections 6 and 7. It remains to treat (III) and (IV).

8.4 (III) Convergent Iteration for the ∞-dimensional Problem

When (54) involves a symmetric positive definite form the matrix $\mathbf{L}$ is symmetric positive definite and, by (81), well-conditioned. Hence simple iterative schemes like gradient or Richardson iterations would converge for any initial guess with some fixed reduction rate. However, in general $\mathbf{L}$ is not definite. The idea is therefore to transform $\mathbf{LU} = \mathbf{F}$ again into an equivalent system

$$\mathbf{LU} = \mathbf{F} \iff \mathbf{Mp} = \mathbf{G}$$

with possibly different coordinates $\mathbf{p}$, such that one still has

$$c_M \|\mathbf{q}\|_{\ell_2} \leq \|\mathbf{Mq}\|_{\ell_2} \leq C_M \|\mathbf{q}\|_{\ell_2}, \quad \mathbf{q} \in \ell_2, \tag{88}$$

while in addition there exists some relaxation weight ω with

$$\|\mathbf{I} - \omega\mathbf{M}\|_{\ell_2 \to \ell_2} \leq \rho < 1. \tag{89}$$

Thus simple iterations of the form

$$\mathbf{p}^{n+1} = \mathbf{p}^n + \omega(\mathbf{G} - \mathbf{Mp}^n) \tag{90}$$

would still converge with a reduction rate $\leq \rho$ per step.

Of course, one could think of more sophisticated iterations with better convergence properties. We will confine the discussion though in essence to (90) for two reasons. First, it makes things much more transparent. Second the ideal iteration (90) will eventually have to be perturbed in practical realizations. There is always a risk of loosing superior convergence properties when performing the applications of the involved operators only approximately. Thus economizing on the application of operators, offers a chance to accomplish the reduction rate offered by (90) at *minimal cost.* In fact, it will be shown later that this leads to schemes that achieve the target accuracy at asymptotically minimal cost.

In the following we discuss two choices of $\mathbf{M}$ for which the above program can be carried through. The first one works for *any* well-posed problem, the second one offers an alternative for saddle point problems. For further options in the latter case, see [27].

Two choices of M

Following [44, 27] we put

$$\mathbf{M} := \mathbf{L}^T\mathbf{L}, \quad \mathbf{G} := \mathbf{L}^T\mathbf{F}, \quad \mathbf{p} = \mathbf{U}. \tag{91}$$

As pointed out in Theorem 6, one then has

$$\mathbf{MU} = \mathbf{G} \quad \Longleftrightarrow \quad \|\mathcal{L}U - F\|_{\mathcal{H}'} = \min_{V \in \mathcal{H}} \|\mathcal{L}V - F\|_{\mathcal{H}'}, \tag{92}$$

so that (89) can be realized whenever (57) and (79) hold. Of course, the condition number still gets squared so that the quantitative behavior of the iteration (90) may in this case be rather poor.

We will therefore discuss an alternative for the class of saddle point problems, see Section 6. In this case, it will turn out that one can take $\mathbf{M}$ as the *Schur complement*

$$\mathbf{M} = \mathbf{B}\boldsymbol{A}^{-1}\mathbf{B}^T. \tag{93}$$

The interpretation of (89) for the least squares formulation is clear, because the application of $\mathbf{M}$ means to apply successively $\mathbf{L}$ and then $\mathbf{L}^T$. The question then remains how to approximate the application of $\mathbf{L}$ and $\mathbf{L}^T$, which will be discussed later.

When $\mathbf{M}$ is the Schur complement, things are less clear even for the idealized situation because of the occurrence of $\boldsymbol{A}^{-1}$. We shall therefore point out next how (89) is to be understood even in the idealized infinite dimensional case.

The application of the Schur complement will be facilitated with the aid of an *Uzawa iteration*, see [37]. To explain this, we need some preparations and recall that in the saddle point case we have $\mathcal{H} = X \times M$ (cf. Section 6.4), and that we need a wavelet basis for each component space

$$X \leftrightarrow \Psi_X \qquad M \leftrightarrow \Psi_M.$$

The norm equivalences (78) then read

$$c_X \|\mathbf{v}\|_{\ell_2(\mathcal{J}_X)} \leq \|\mathbf{v}^T \mathbf{D}_X^{-1} \Psi_X\|_X \leq C_X \|\mathbf{v}\|_{\ell_2(\mathcal{J}_X)}. \tag{94}$$

and

$$c_M \|\mathbf{q}\|_{\ell_2(\mathcal{J}_M)} \leq \|\mathbf{q}^T \mathbf{D}_M^{-1} \Psi_M\|_M \leq C_M \|\mathbf{q}\|_{\ell_2(\mathcal{J}_M)}. \tag{95}$$

Specifically, in the case of the Stokes problem one can take the scaling weights $(\mathbf{D}_X)_{\lambda,\lambda} = 2^{|\lambda|}$, $(\mathbf{D}_M)_{\lambda,\lambda} = 1$.

Setting,

$$\boldsymbol{A} := \mathbf{D}_X^{-1} a(\Psi_X, \Psi_X) \mathbf{D}_X^{-1}, \quad \mathbf{B} := \mathbf{D}_M^{-1} b(\Psi_M, \Psi_X) \mathbf{D}_X^{-1},$$

and

$$\mathbf{f} := \mathbf{D}_X^{-1} \langle \Psi_X, f \rangle, \quad \mathbf{g} := \mathbf{D}_M^{-1} \langle \Psi_M, g \rangle,$$

(67) is equivalent to

$$\mathbf{L}\mathbf{U} = \mathbf{F} \quad \Longleftrightarrow \quad \underbrace{\begin{pmatrix} \boldsymbol{A} & \mathbf{B}^T \\ \mathbf{B} & \mathbf{0} \end{pmatrix}}_{\mathbf{L}} \underbrace{\begin{pmatrix} \mathbf{u} \\ \mathbf{p} \end{pmatrix}}_{\mathbf{U}} = \underbrace{\begin{pmatrix} \mathbf{f} \\ \mathbf{g} \end{pmatrix}}_{\mathbf{F}}. \tag{96}$$

Moreover, under the assumptions (69), (70) together with (94), (95) the mapping property

$$c_L \|\mathbf{V}\|_{\ell_2} \leq \|\mathbf{L}\mathbf{V}\|_{\ell_2} \leq C_L \|\mathbf{V}\|_{\ell_2}, \quad \mathbf{V} \in \ell_2, \tag{97}$$

holds, see Theorem 5

One actually has to be somewhat careful with (94), which expresses that $\boldsymbol{A}$ is invertible on all of ℓ_2 or, in other words, that A is invertible on all of X. Recall from (69) that this is neither necessary nor generally the case. However, whenever the saddle point problem (67) is well-posed, one can show that for some suitable $c > 0$ the matrix $\hat{\boldsymbol{A}} := \boldsymbol{A} + c\mathbf{B}^T\mathbf{B}$ is invertible on all of ℓ_2 and satisfies (94). One can then replace (96) by an equivalent system (with adjusted right hand side data) so that (MP) is valid and the Schur complement is well defined. Therefore we will assume in the following that either A is invertible on all of X, or that the above precaution has been been used, leading to a matrix $\hat{\boldsymbol{A}}$, henceforth again denoted by $\boldsymbol{A}$, so that (93) makes sense.

Thus we can use block elimination and observe that (96) is equivalent to

$$\begin{cases} \mathbf{M}\mathbf{p} := \mathbf{B}\boldsymbol{A}^{-1}\mathbf{B}^T\mathbf{p} = \mathbf{B}\boldsymbol{A}^{-1}\mathbf{f} - \mathbf{g} =: \mathbf{G} \\ \boldsymbol{A}\mathbf{u} \qquad\qquad\qquad\quad = \mathbf{f} - \mathbf{B}^T\mathbf{p}. \end{cases}$$

On account of (97), we know that

$$\mathbf{M} := \mathbf{B}\boldsymbol{A}^{-1}\mathbf{B}^T : \ell_2(\mathcal{J}_M) \to \ell_2(\mathcal{J}_M), \quad \|\mathbf{M}\mathbf{q}\|_{\ell_2(\mathcal{J}_M)} \sim \|\mathbf{q}\|_{\ell_2(\mathcal{J}_M)}. \tag{98}$$

Therefore there exists a positive relaxation parameter ω such that a fixed point iteration (or a gradient iteration) based on the identity

$$\mathbf{p} = \mathbf{p} + \omega\left((\mathbf{B}\boldsymbol{A}^{-1}\mathbf{f} - \mathbf{g}) - \mathbf{M}\mathbf{p}\right)$$
$$= \mathbf{p} + \omega\left(\mathbf{B}\underbrace{\boldsymbol{A}^{-1}(\mathbf{f} - \mathbf{B}^T\mathbf{p})}_{=\mathbf{u}} - \mathbf{g}\right) = \mathbf{p} + \omega(\mathbf{B}\mathbf{u} - \mathbf{g})$$

converges with a fixed reduction rate $\rho < 1$. Thus replacing for some iterate $\mathbf{p}^n$ the expression $\boldsymbol{A}^{-1}(\mathbf{f} - \mathbf{B}^T\mathbf{p}^n)$ in this iteration by the solution $\mathbf{u}^n$ of

$$\boldsymbol{A}\mathbf{u}^n = \mathbf{f} - \mathbf{B}^T\mathbf{p}^n, \tag{99}$$

the iteration (90) for $\mathbf{M}$ given by (93), reduces to the simple update

$$\mathbf{p}^{n+1} = \mathbf{p}^n + \omega(\mathbf{B}\mathbf{u}^n - \mathbf{g}), \tag{100}$$

with $\mathbf{u}^n$ from (99). This is the idea of the Uzawa scheme here formulated for the original infinite dimensional problem in wavelet coordinates.

We are now prepared to discuss the last step (IV).

8.5 (IV) Adaptive Application of Operators

For notational simplicity we dispense with distinguishing between the unknowns $\mathbf{U}$ and $\mathbf{p}$ (see Section 8.4) and continue to use $\mathbf{U}$ for the moment to denote the vectors of wavelet coefficients. Our objective is to turn the idealized iteration

$$\mathbf{M}\mathbf{U} = \mathbf{G} \quad \rightsquigarrow \quad \mathbf{U}^{n+1} = \mathbf{U}^n + \omega(\mathbf{G} - \mathbf{M}\mathbf{U}^n)$$

into a practicable version. Neither can we evaluate the generally infinite array $\mathbf{G}$ exactly, nor can we compute $\mathbf{M}\mathbf{U}^n$, even when $\mathbf{U}^n$ has finite support. Thus, we need approximations to these two ingredients, which we will formulate as

Basic Routines:

RHS $[\eta, \mathbf{G}] \to \mathbf{G}_\eta$: such that $\|\mathbf{G} - \mathbf{G}_\eta\|_{\ell_2} \le \eta$;

APPLY $[\eta, \mathbf{M}, \mathbf{V}] \to \mathbf{W}_\eta$: such that $\|\mathbf{M}\mathbf{V} - \mathbf{W}_\eta\|_{\ell_2} \le \eta$;

COARSE $[\eta, \mathbf{W}] \to \bar{\mathbf{W}}_\eta$: such that $\|\mathbf{W} - \bar{\mathbf{W}}_\eta\|_{\ell_2} \le \eta$.

A few comments on these routines are in order. The input to **APPLY** and **COARSE** will always be finitely supported. **COARSE** can then be realized by sorting the coefficients and by adding successively the squares of the entries from small to large, until the target threshold is reached. For details see [26]. It is pointed out in [7] how to avoid log-terms caused by sorting.

The simplest example of **RHS** is encountered in the scalar elliptic case. Then $\mathbf{G} = \mathbf{f}$ consists of the dual wavelet coefficients of the right hand side f. One should then think of computing in a preprocessing step a highly accurate approximation to f in the dual basis along with the corresponding coefficients. The necessary accuracy can be easily related to the target accuracy of the adaptive process. Ordering these finitely many coefficients by size, **RHS** becomes then an application of **COARSE** to this finitely supported array. In more general cases, e.g. in the least squares formulation, **RHS** may take different formats to be discussed for each special case. However, in general it will always involve a combination of the two last routines **APPLY** and **COARSE**.

It is less obvious how to go about the routine **APPLY**, which will be explained later in more detail.

8.6 The Adaptive Algorithm

We will *assume* for the moment that we have the above routines at hand and wish to determine first for which tolerances η the corresponding perturbation of the ideal scheme (90) converges.

SOLVE $[\epsilon, \mathbf{M}, \mathbf{G}] \to \bar{\mathbf{U}}(\epsilon)$

(i) Set $\bar{\mathbf{U}}^0 = \mathbf{0}, \quad \epsilon_0 := c_M^{-1}\|\mathbf{G}\|_{\ell_2}, \quad j = 0.$
(ii) If $\epsilon_j \le \epsilon$, *stop* $\bar{\mathbf{U}}^j \to \mathbf{U}(\epsilon)$. *Else* $\mathbf{V}^0 := \bar{\mathbf{U}}^j$.

(ii.1) *For* $l = 0, \dots, K-1$:

RHS $[\rho^l \epsilon_j, \mathbf{G}] \to \mathbf{G}_l$; **APPLY** $[\rho^l \epsilon_j, \mathbf{M}, \mathbf{V}^l] \to \mathbf{W}^l$; *and set*

$$\mathbf{V}^{l+1} := \mathbf{V}^l + \alpha(\mathbf{G}_l - \mathbf{W}^l).$$

(ii.2) **COARSE** $[\mathbf{V}^K, 2\epsilon_j/5] \to \bar{\mathbf{U}}^{j+1}, \quad \epsilon_{j+1} := \epsilon_j/2,\ j+1 \to j$ *go to* **(ii)**.

Here ρ is the reduction rate from (89) and K depends only on the constants in (79), (88) and (57). In fact, it can be shown that, based on these constants and the reduction rate ρ in (89), there exists a uniformly bounded K so that $\|\mathbf{U} - \mathbf{V}^K\|_{\ell_2} \le \epsilon_j/10$. The coarsening step (ii.2) leads then to the following estimate.

Proposition 12 *The approximations* $\bar{\mathbf{U}}^j$ *satisfy*

$$\|\mathbf{U} - \bar{\mathbf{U}}^j\|_{\ell_2} \le \epsilon_j, \quad j \in \mathbb{N}. \tag{101}$$

Thus any target accuracy is met after finitely many steps. Note that the accuracy tolerances are at each stage comparable to the current accuracy, which will be important for later complexity estimates.

The above scheme should be viewed as the simplest example of a perturbed representation of an iteration for the infinite dimensional problem. Several alternatives come to mind. Instead of applying always K steps (ii.1), one can monitor the approximate residual for possible earlier termination. Furthermore, the fixed relaxation parameter α can be replaced by a stage dependent parameter α_j resulting from a line search in a *gradient iteration*. Finally, one could resort to (approximate) conjugate gradient iterations. To minimize technicalities we will stick, however, with the above simpler Richardson scheme.

The central question now is how to realize the basic routines in practice and what is their complexity.

8.7 Ideal Bench Mark – Best N-Term Approximation

We wish to compare the performance of the above algorithm with what could be achieve *ideally*, namely with the work/accuracy balance of the best N-term approximation, recall (86). Since the relevant domain is just ℓ_2 the following version matters.

$$\sigma_{N,\ell_2}(\mathbf{V}) := \|\mathbf{V} - \mathbf{V}_N\|_{\ell_2} = \min_{\#\mathrm{supp}\,\mathbf{W} \le N} \|\mathbf{V} - \mathbf{W}\|_{\ell_2}. \tag{102}$$

Due to the norm equivalences (79), one has

$$\sigma_{N,\ell_2}(\mathbf{V}) \sim \inf_{\mathbf{W},\#\mathrm{supp}\,\mathbf{W} \le N} \|V - \mathbf{W}^T \mathbf{D}^{-1} \Psi\|_{\mathcal{H}} = \sigma_{N,\mathcal{H}}(V), \tag{103}$$

i.e., the best N-term approximation in the computational domain ℓ_2 corresponds directly to the best N-term approximation in the energy norm.

The following interrelation between best N-term approximation and coarsening or *thresholding* sheds some light on the role of **COARSE** in step (ii.2) of algorithm **SOLVE**, see [26].

Remark 13 *Suppose the the finitely supported vector* $\mathbf{w}$ *satisfies* $\|\mathbf{v} - \mathbf{w}\|_{\ell_2} \le \eta/5$. *Clearly* $\bar{\mathbf{w}}_\eta :=$ **COARSE** $[\mathbf{w}, 4\eta/5]$ *still satisfies* $\|\mathbf{v} - \bar{\mathbf{w}}\|_{\ell_2} \le \eta$. *Moreover, whenever* $\|\mathbf{v} - \mathbf{v}_N\|_{\ell_2} \lesssim N^{-s}$ *one has*

$$\#\mathrm{supp}\,\bar{\mathbf{w}}_\eta \lesssim \eta^{-1/s}, \quad \|\mathbf{v} - \bar{\mathbf{w}}_\eta\|_{\ell_2} \le \eta. \tag{104}$$

Thus the application of **COARSE** pulls a current approximation towards the best N-term approximation.

8.8 Compressible Matrices

We turn now to the routine **APPLY**. At this point the cancellation properties (12) comes into play. As indicated in Section 2.2 wavelet representations of many operators are quasi-sparse. In the present context the following quantification of sparsity or *compressibility* is appropriate [26].

A matrix $\mathbf{C}$ is said to be s^*-compressible – $\mathbf{C} \in \mathcal{C}_{s^*}$ – if for any $0 < s < s^*$ and every $j \in \mathbb{N}$, there exists a matrix $\mathbf{C}_j$ obtained by replacing all but the order of $\alpha_j 2^j$ ($\sum_j \alpha_j < \infty$) entries per row and column in $\mathbf{C}$ by zero, while still

$$\|\mathbf{C} - \mathbf{C}_j\| \le \alpha_j 2^{-js}, \quad j \in \mathbb{N}, \quad \sum_j \alpha_j < \infty. \tag{105}$$

As mentioned above, one can use the cancellation properties (CP) to confirm the following claim.

Remark 14 *The scaled wavelet representations* $\mathbf{L}_{i,l}$ *in the above examples belong to* $\mathcal{C}_{s^*}$ *for some* $s^* = s^*(\mathcal{L}, \Psi) > 0$.

8.9 Fast Approximate Matrix/Vector Multiplication

The key is that for compressible matrices one can devise approximate application schemes that exhibit in a certain range an asymptotically optimal work/accuracy balance. To this end, consider for simplicity the scalar case and abbreviate for any finitely supported $\mathbf{v}$ the best 2^j-term approximations as $\mathbf{v}_{[j]} := \mathbf{v}_{2^j}$ and define

$$\mathbf{w}_j := \boldsymbol{A}_j \mathbf{v}_{[0]} + \boldsymbol{A}_{j-1}(\mathbf{v}_{[1]} - \mathbf{v}_{[0]}) + \cdots + \boldsymbol{A}_0(\mathbf{v}_{[j]} - \mathbf{v}_{[j-1]}), \tag{106}$$

as an approximation to $\boldsymbol{A}\mathbf{v}$. In fact, the triangle inequality together with the above compression estimates yield

$$\|\boldsymbol{A}\mathbf{v} - \mathbf{w}_j\|_{\ell_2} \leq c \underbrace{\|\mathbf{v} - \mathbf{v}_{[j]}\|_{\ell_2}}_{\sigma_{2^j,\ell_2}(\mathbf{v})} + \sum_{l=0}^{j} \alpha_l 2^{-ls} \underbrace{\|\mathbf{v}_{[j-l]} - \mathbf{v}_{[j-l-1]}\|_{\ell_2}}_{\lesssim\ \sigma_{2^{j-l-1},\ell_2}(\mathbf{v})}. \tag{107}$$

One can now exploit the *a-posteriori* information offered by the quantities $\sigma_{2^{j-l-1},\ell_2}(\mathbf{v})$ to choose the smallest j for which the right hand side of (107) is smaller than a given target accuracy η. Since the sum is finite for each finitely supported input $\mathbf{v}$ such a j does indeed exist. This leads to a concrete multiplication scheme

$$\textbf{MULT}\,[\eta, \boldsymbol{A}, \mathbf{v}] \to \mathbf{w}_\eta \quad \text{s.t.:} \quad \|\boldsymbol{A}\mathbf{v} - \mathbf{w}_\eta\| \leq \eta,$$

which is analyzed in [26] and implemented in [8]. The main result can be formulated as follows [26].

Theorem 7. *If* $\boldsymbol{A} \in \mathcal{C}_{s^*}$ *and* $\|\mathbf{v} - \mathbf{v}_N\|_{\ell_2} \lesssim N^{-s}$ *for* $s < s^*$, *then*

$$\#\mathrm{supp}\, \mathbf{w}_\eta \lesssim \eta^{-1/s}, \quad \#\textbf{flops} \lesssim \#\mathrm{supp}\, \mathbf{v} + \eta^{-1/s}.$$

Thus, **MULT** has in some range an asymptotically optimal work/accuracy balance. In fact, it is pointed out in [7] that an original logarithmic factor, due to sorting operations can be avoided.

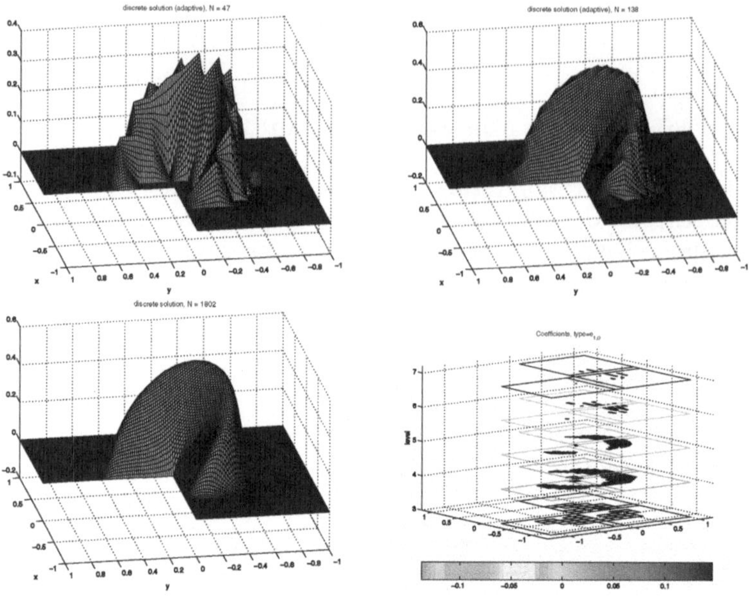

Fig. 6. Adaptive solution of the Poisson problem on a L-shaped domain

Note that whenever the bilinear form in (54) is coercive, one can take $\mathbf{M} = \mathbf{L}$ to be just the wavelet representation of $\mathcal{L}$, which is typically s^*-compressible for some positive s^*. In this case one can take **APPLY** = **MULT**. Again Poisson's equation, i.e., (58) with $a(x)$ the identity matrix and $k(x) = 0$, may serve as an example. Figure 6 illustrates the progressive formation of the solution when $\Omega \subset \mathbb{R}^2$ is an L-shaped domain. The lower right part displays the position of the significant coefficients of the approximate solution at the final stage. The reentrant corner causes a singularity in spite of a smooth right hand side. In the example shown in Figure 6 the right hand side is chosen so as to create the strongest possible singularity for that type of problems, see [8] for more details. The role of the regularity of the solution with regard to the computational complexity will be addressed later.

In general $\mathbf{M}$ will differ from $\mathbf{L}$ and we will indicate briefly how to choose the scheme **APPLY** for the two choices of $\mathbf{M}$ mentioned above.

The simplest case is the least squares formulation $\mathbf{M} = \mathbf{L}^T\mathbf{L}$, $\mathbf{G} = \mathbf{L}^T\mathbf{F}$, $\mathbf{p} = \mathbf{U}$.

$$\mathbf{RHS}_{ls}[\eta, \mathbf{G}] := \mathbf{MULT}\left[\tfrac{\eta}{2}, \mathbf{L}^T, \mathbf{COARSE}\left[\tfrac{\eta}{2C_L}, \mathbf{F}\right]\right]$$

$$\mathbf{APPLY}_{ls}[\eta, \mathbf{M}, \mathbf{V}] := \mathbf{MULT}\left[\tfrac{\eta}{2}, \mathbf{L}^T, \mathbf{MULT}\left[\tfrac{\eta}{2C_L}, \mathbf{L}, \mathbf{V}\right]\right]$$

In this case the routines **RHS** and **APPLY** are compositions of **COARSE** and **MULT**.

The situation is more involved when employing Uzawa iterations for saddle point problems as explained next. The reader is referred to [36] for details.

8.10 Application Through Uzawa Iteration

As indicated before, the application of the Schur complement $\mathbf{M}$ (98) can be based on Uzawa iterations. To explain this, recall first the

IDEAL UZAWA: *Given any* $\mathbf{p}_0 \in \ell_2(\mathcal{J}_M)$, *compute for* $i = 1, 2, \ldots$

$$\begin{aligned} \boldsymbol{A}\mathbf{u}_i &= \mathbf{f} - \mathbf{B}^T\mathbf{p}_{i-1}, \qquad (\text{set } \ \boldsymbol{R} := \langle \tilde{\Psi}_M, \tilde{\Psi}_M \rangle) \\ \mathbf{p}_i &= \mathbf{p}_{i-1} + \omega \boldsymbol{R}(\mathbf{B}\mathbf{u}_i - \mathbf{g}), \qquad \|\mathbf{I} - \omega \boldsymbol{R}\mathbf{S}\|_{\ell_2 \to \ell_2} \leq \rho < 1 \end{aligned} \tag{108}$$

A few comments on the role of $\boldsymbol{R}$ are in order. The operator B in (68) maps X to M' while p belongs to M. In principle, this does not matter since the wavelet transform maps all spaces to ℓ_2. However, as in the case of the Stokes problem, M may be a closed subspace (with finite codimension) of some larger Hilbert space that permits a wavelet characterization. In the Stokes problem $L_{2,0}(\Omega)$ is obtained by factoring out constants. Due to the vanishing moments of wavelets, these constants have to be removed only from the finite dimensional coarse scale space. But the corresponding linear relations on the coefficients are in general different from their counterparts for the dual basis $\tilde{\Psi}_M$. Since the operators are not applied exactly a correction through the application of $\boldsymbol{R}$ that maps the vectors into the "right domain" turns out to be necessary, see [36].

The application of $\mathbf{M}$ consists then of solving the elliptic problem in (108) approximately by invoking an elliptic version of **SOLVE** (with **APPLY** = **MULT**) called **ELLSOLVE**. The application of $\mathbf{B}$, $\mathbf{B}^T$ and $\boldsymbol{R}$ in turn is facilitated by **MULT**. See [36] for details and corresponding precise tolerances.

8.11 Main Result – Convergence/Complexity

We return now to the general problem (54). In the following it is understood that the scheme **APPLY** is based on either one of the above versions. Moreover, full accessibility of the right hand side data is assumed. Finally, we will assume that the entries of $\mathbf{L}$ can be computed on average at unit cost. This is justified, for instance, when dealing with constant coefficient differential operators. In general, this is a delicate task that often motivates to resort to the so called *nonstandard form*. However, it can be shown that also in more realistic situations this assumption can be realized, see also [7, 9]. The main result can then be formulated as follows.

Theorem 8. *Assume that* $\mathbf{L} \in \mathcal{C}_{s^*}$ *for some* $s^* = s^*(\mathcal{L}, \Psi) > 0$. *Then if the exact solution* $U = \mathbf{U}^T \mathbf{D}^{-1} \Psi$ *of (54) satisfies for some* $s < s^*$

$$\inf_{\#\mathrm{supp}\,\mathbf{V} \leq N} \|U - \mathbf{V}^T \mathbf{D}^{-1} \Psi\|_{\mathcal{H}} \lesssim N^{-s},$$

then, for any $\epsilon > 0$, *the approximations* $\bar{\mathbf{U}}(\epsilon)$ *produced by* **SOLVE** *satisfy*

$$\|U - \bar{\mathbf{U}}(\epsilon)^T \mathbf{D}^{-1} \Psi\|_{\mathcal{H}} \lesssim \epsilon$$

and

$$\#\mathrm{supp}\,\bar{\mathbf{U}}(\epsilon), \text{ comp. work } \lesssim \epsilon^{-1/s}.$$

It should be noted that, aside from the complexity aspect, the adaptive scheme has the interesting effect, that compatibility conditions like the *LBB condition* become void. Roughly speaking, the adaptive application of the (full infinite dimensional) operators within the right accuracy tolerances inherits at each stage enough of the stability of the infinite dimensional problem. At no stage a fixed trial space is chosen that requires taking care of such constraints.

8.12 Some Ingredients of the Proof of Theorem 8

We shall discuss next some typical ingredients from *nonlinear approximation* entering the proof of Theorem 8, see [24, 51, 54] for more details. Recall that typical approximation errors for spaces arising from spatial refinements decay like powers of the mesh size and thus like negative powers of the number of degrees of freedom. In view of (103), it is therefore of interest to understand the structure of those sequences $\mathbf{v} \in \ell_2$ whose best N-term approximation decays like N^{-s} for some $s > 0$. Since best N-term approximation in ℓ_2 is realized by retaining simply the largest N coefficients, it is directly related to *thresholding.*

Thresholding

One way to describe sequences that are *sparse* in the sense that $\sigma_{N,\ell_2}(\mathbf{v})$ decays fast, is to control the number of terms exceeding any given threshold. To this end, define the thresholding operator

$$\mathcal{T}_\eta \mathbf{v} := \begin{cases} v_\lambda, & |v_\lambda| \geq \eta, \\ 0, & |v_\lambda| < \eta, \end{cases}$$

and set for some $\tau < 2$

$$\ell_\tau^w := \left\{ \mathbf{v} \in \ell_2 : N(\mathbf{v}, \eta) := \#\mathrm{supp}\,\mathcal{T}_\eta \mathbf{v} \leq C_v \eta^{-\tau} \right\}. \tag{109}$$

In fact, for $\tau \geq 2$ the condition $\#\mathrm{supp}\,\mathcal{T}_\eta \mathbf{v} \leq C_v \eta^{-\tau}$ is satisfied for all $\mathbf{v} \in \ell_2$.

For sequences $\mathbf{v}$ in ℓ_τ^w the error produced by thresholding is easily determined, noting that for $\mathbf{v} \in \ell_\tau^w$

$$\begin{aligned}\|\mathbf{v} - \mathcal{T}_\eta \mathbf{v}\|_{\ell_2}^2 &= \sum_{l=0}^{\infty} \sum_{2^{-l-1}\eta \le |v_\lambda| < 2^{-l}\eta} |v_\lambda|^2 \le C_v \sum_{l=0}^{\infty} (2^{-l}\eta)^2 (2^{-l-1}\eta)^{-\tau} \\ &= \frac{4C_v}{2^{2-\tau} - 1}\eta^{2-\tau}. \end{aligned} \tag{110}$$

To understand the nature of the constant C_v appearing in the definition of ℓ_τ^w, consider the *decreasing rearrangement* $(v_n^*)_{n\in\mathbb{N}}$ of $\mathbf{v}$ defined by $v_{n+1}^* \le v_n^*$, $v_n^* = |v_{\lambda_n}|$. By definition one has

$$v_{N(\mathbf{v},\eta)+1}^* N(\mathbf{v},\eta)^{1/\tau} \le \eta N(\mathbf{v},\eta)^{1/\tau} \le C_v^{1/\tau}.$$

Thus defining

$$C_v^{1/\tau} = \sup_{n\in\mathbb{N}} n^{1/\tau} v_{1+n}^* =: |\mathbf{v}|_{\ell_\tau^w}, \tag{111}$$

we see that

$$\|\mathbf{v}\|_{\ell_\tau^w} := \|\mathbf{v}\|_{\ell_2} + |\mathbf{v}|_{\ell_\tau^w} \tag{112}$$

is a (quasi-) norm for ℓ_τ^w. The next observation is that ℓ_τ^w is very close to ℓ_τ. In fact

$$n^{1/\tau} v_{n+1}^* \le (n(v_n^*)^\tau)^{1/\tau} \le \left(\sum_{j\le n} (v_j^*)^\tau\right)^{1/\tau} \le \|\mathbf{v}\|_{\ell_\tau},$$

so that

$$\ell_\tau \subset \ell_\tau^w \subset \ell_{\tau+\epsilon} \subset \ell_2, \quad \tau < \tau + \epsilon < 2. \tag{113}$$

The estimate (110) can be used to establish the following facts, that will serve as prerequisites for Remark 13, see [26].

Lemma 15. *Suppose that $\mathbf{v} \in \ell_\tau^w$ for some $0 < \tau < 2$ and that $\mathbf{w} \in \ell_2$ satisfies*

$$\|\mathbf{v} - \mathbf{w}\|_{\ell_2} \le \epsilon \quad \textit{for some} \quad \epsilon > 0.$$

Then one has for any $\eta > 0$

$$\|\mathbf{v} - \mathcal{T}_\eta \mathbf{w}\|_{\ell_2} \le 2\epsilon + \bar{C}\|\mathbf{v}\|_{\ell_\tau^w}^{\tau/2} \eta^{1-\tau/2}, \quad \#\mathrm{supp}\, \mathcal{T}_\eta \mathbf{w} \le \frac{4\epsilon^2}{\eta^2} + 4\bar{C}\|\mathbf{v}\|_{\ell_\tau^w}^\tau \eta^{-\tau},$$

where $\bar{C}$ depends only on τ as $\tau \to 0$.

This can be used to prove Remark 13, see [26].

Characterization of the Rates N^{-s}

We now turn to the characterization of those sequences whose best N-Term approximation rates decay like N^{-s} and recall that

$$\sigma_{N,\ell_2}(\mathbf{v}) := \|\mathbf{v} - \mathbf{v}_N\|_{\ell_2} = \min_{\#\mathrm{supp}\, \mathbf{w} \le N} \|\mathbf{v} - \mathbf{w}\|_{\ell_2}.$$

Proposition 16 *Let $\frac{1}{\tau} = s + \frac{1}{2}$. Then* $\mathbf{v} \in \ell_\tau^w$ *if and only if* $\sigma_{N,\ell_2}(\mathbf{v}) \lesssim N^{-s}$ *and*

$$\|\mathbf{v} - \mathbf{v}_N\|_{\ell_2} \lesssim N^{-s}\|\mathbf{v}\|_{\ell_\tau^w}.$$

Proof:

$$\begin{aligned}\sigma_{N,\ell_2}(\mathbf{v})^2 &= \sum_{n>N}(v_n^*)^2 \le \left(\sum_{n\ge N} n^{-2/\tau}\right)|\mathbf{v}|^2_{\ell_\tau^w} \le C\,N^{1-\frac{2}{\tau}}|\mathbf{v}|^2_{\ell_\tau^w}\\ &= C\,N^{-2s}|\mathbf{v}|^2_{\ell_\tau^w}.\end{aligned}$$

Conversely,

$$N|v_{2N}^*|^2 \le \sum_{N<n\le 2N}|v_n^*|^2 \le \sigma_{N,\ell_2}(\mathbf{v})^2 \le C\,N^{-2s},$$

which means that

$$|v_{2N}^*| \le C\,N^{-(s+1/2)} = C\,N^{-1/\tau},$$

and hence $\mathbf{v} \in \ell_\tau^w$, which completes the proof. □

Key Complexity Bounds

The above preliminaries can be used to show that the basic routines in **SOLVE** exhibit under certain circumstances optimal work/accuracy balances. In fact, counting the terms in (106) and taking the characterization of $\sigma_{N,\ell_2}(\mathbf{v}) = \mathcal{O}(N^{-s})$ given in Proposition 16 into account, leads to the following fact, which implies Theorem 7 stated above.

Proposition 17 *Suppose that* $\mathbf{C} \in \mathcal{C}_{s^*}$ *and let* $\frac{1}{\tau} = s + \frac{1}{2}$, $s < s^*$. *Then* $\mathbf{w}_\eta = \mathbf{MULT}\,[\eta, \mathbf{C}, \mathbf{v}]$ *satisfies for any finitely supported input* $\mathbf{v}$

- $\|\mathbf{w}_\eta\|_{\ell_\tau^w} \lesssim \|\mathbf{v}\|_{\ell_\tau^w}$
- $\#\text{flops} \lesssim \#\text{supp}\,\mathbf{v} + \|\mathbf{v}\|_{\ell_\tau^w}^{1/s}\eta^{-1/s}$, $\#\text{supp}\,\mathbf{w}_\eta \lesssim \|\mathbf{v}\|_{\ell_\tau^w}^{1/s}\eta^{-1/s}$,

which indeed confirms the optimal balance: *accuracy* $\eta \leftrightarrow$ *cost* $\eta^{-1/s}$ *for* $s < s^*$.

This, in turn, is one of the main ingredients of a proof of the following result [27].

Theorem 9. *Let* $\mathbf{L} \in \mathcal{C}_{s^*}$ *and suppose that* $\mathbf{U} \in \ell_\tau^w$ *for* $\frac{1}{\tau} = s + \frac{1}{2}$, $s < s^*$. *Then in all the above cases the output* $\mathbf{G}_\eta$ *of the right hand side scheme* **RHS** $[\eta, \mathbf{G}]$ *satisfies*

(1) $\|\mathbf{G}_\eta\|_{\ell_\tau^w} \lesssim \|\mathbf{G}\|_{\ell_\tau^w} \lesssim \|\mathbf{U}\|_{\ell_\tau^w}$;
(2) $\#\text{flops} \sim \#\text{supp}\,\mathbf{G}_\eta \lesssim \|\mathbf{G}\|_{\ell_\tau^w}^{1/s}\eta^{-1/s}$.

Moreover, for **APPLY** $\in$ {**MULT**, **APPLY**$_{ls}$, **APPLY**$_{Uz}$} *the output* $\mathbf{W}_\eta$ *of* **APPLY** $[\eta, \mathbf{M}, \mathbf{V}]$ *satisfies for* $s < s^*$ *and any finitely supported input* $\mathbf{V}$*:*

(3) $\|\mathbf{W}_\eta\|_{\ell^w_\tau} \lesssim \|\mathbf{V}\|_{\ell^w_\tau}$*;*

(4) $\#\text{flops} \lesssim \#\text{supp}\,\mathbf{V} + \|\mathbf{V}\|_{\ell^w_\tau}^{1/s}\eta^{-1/s}$, $\#\text{supp}\,\mathbf{W}_\eta \lesssim \|\mathbf{V}\|_{\ell^w_\tau}^{1/s}\eta^{-1/s}$.

Thus, according to Theorem 8, **APPLY** $[\epsilon, \mathbf{M}^{-1}, \mathbf{G}] :=$ **SOLVE** $[\epsilon, \mathbf{M}, \mathbf{G}]$ exibits the same work/accuracy balance as its ingredients. In fact, the proof of Theorem 8 is based on Theorem 9 above.

Compressibility Criteria

We conclude this section with some sufficient conditions for a wavelet representation $\mathbf{L}$ to be compressible. To this end, recall that in all the examples considered before, the operator $\mathcal{L}$ is either *local* or of the form

$$(\mathcal{L}u)(x) = \int_\Gamma K(x,y)u(y)\,d\Gamma_y,$$

where

$$\left|\partial_x^\alpha \partial_y^\beta K(x,y)\right| \lesssim \text{dist}(x,y)^{-(d+2t+|\alpha|+|\beta|)}. \tag{114}$$

Results of the following type can be found e.g. in [45, 66].

Theorem 10. *Suppose that* $\mathcal{L}$ *has order* $2t$ *and satisfies for some* $r > 0$

$$\|\mathcal{L}v\|_{H^{-t+a}} \lesssim \|v\|_{H^{t+a}}, \quad v \in H^{t+a}, 0 \le |a| \le r.$$

Assume that $\mathbf{D}^{-s}\Psi$ *is a Riesz-basis for* H^s *for* $-\tilde{\gamma} < s < \gamma$ *(16) and has cancellation properties (CP) (12) of order* $\tilde{m}$*. Then for any* $0 < \sigma \le \min\{r, d/2 + \tilde{m} + t\}$, $t + \sigma < \gamma$, $t - \sigma > -\tilde{\gamma}$*, one has*

$$2^{-(|\lambda'|+|\lambda|)t}|\langle\psi_\lambda, \mathcal{L}\psi_{\lambda'}\rangle| \lesssim \frac{2^{-||\lambda|-|\lambda'||\sigma}}{(1 + 2^{\min(|\lambda|,|\lambda'|)}\,\text{dist}(\Omega_\lambda, \Omega_{\lambda'}))^{d+2\tilde{m}+2t}}. \tag{115}$$

Thus the entries of the wavelet representation of operators of the above type exhibit a polynomial spatial decay, depending on the order of cancellation properties, and an exponential scalewise decay, depending on the regularity of the wavelets.

Since estimates of this type are also basic for matrix compression techniques, in connection with boundary integral equations, we give a

Sketch of the argument: (See ([45, 66, 67] for more details). Let again $\Omega_\lambda := \text{supp}\,\psi_\lambda$. One distinguishes two cases. Suppose first that $\text{dist}(\Omega_\lambda, \Omega_{\lambda'}) \gtrsim 2^{-\min(|\lambda|,|\lambda'|)}$. Since then x stays away from y the kernel is smooth. In contrast to the nonstandard form one can then apply (CP) (see (12)) of order $\tilde{m}$ to the kernel

$$\int_\Gamma \int_\Gamma K(x,y)\psi_\lambda \psi_{\lambda'} dx dy = \langle K, \psi_\lambda \otimes \psi_{\lambda'} \rangle$$

successively with respect to *both* wavelets. The argument is similar to that in Section 2.2 and one eventually obtains

$$|\langle \mathcal{L}\psi_{\lambda'}, \psi_\lambda \rangle| \lesssim \frac{2^{-(|\lambda|+|\lambda'|)(d/2+\tilde{m})}}{(\mathrm{dist}(\Omega_\lambda, \Omega_{\lambda'}))^{d+2\tilde{m}+2t}}. \tag{116}$$

If on the other hand $\mathrm{dist}(\Omega_\lambda, \Omega_{\lambda'}) \lesssim 2^{-\min(|\lambda|,|\lambda'|)}$ we follow [35] and use the *continuity* of $\mathcal{L}$ together with (NE) (see (16)) to conclude that

$$\|\mathcal{L}v\|_{H^{-t+s}} \lesssim \|v\|_{H^{t+s}}, \quad v \in H^{t+s}, 0 \le |s| \le \tau. \tag{117}$$

Assuming without loss of generality that $|\lambda| > |\lambda'|$ and employing the Schwarz inequality, yields

$$|\langle \mathcal{L}\psi_{\lambda'}, \psi_\lambda \rangle| \le \|\mathcal{L}\psi_{\lambda'}\|_{H^{-t+\sigma}} \|\psi_\lambda\|_{H^{t-\sigma}} \lesssim \|\psi_{\lambda'}\|_{H^{t+\sigma}} \|\psi_\lambda\|_{H^{t-\sigma}}.$$

Under the above assumptions on σ one obtains

$$|\langle \mathcal{L}\psi_{\lambda'}, \psi_\lambda \rangle| \le 2^{t(|\lambda|+|\lambda'|)} 2^{\sigma(|\lambda'|-|\lambda|)}.$$

This completes the proof. □

The estimates for wavelets with overlapping supports can actually be refined using the so called *second compression.* This allows one to remove log-factors in matrix compression estimates arising in the treatment of boundary integral operators, see [62, 68].

Estimates of the type (115) combined with the Schur Lemma lead to the following result [26].

Proposition 18 *Suppose that*

$$|\mathbf{C}_{\lambda,\nu}| \lesssim \frac{2^{-\sigma||\lambda|-|\nu||}}{(1+d(\lambda,\nu))^\beta}, \quad d(\lambda,\nu) := 2^{\min\{|\lambda|,|\nu|\}} \mathrm{dist}\,(\Omega_\lambda, \Omega_\nu)$$

and define

$$s^* := \min\left\{\frac{\sigma}{d} - \frac{1}{2}, \frac{\beta}{d} - 1\right\}.$$

Then $\mathbf{C} \in \mathcal{C}_{s^*}$.

As mentioned above, the proof is based on the

Schur Lemma: *If for some* $C < \infty$ *and any positive sequence* $\{\omega_\lambda\}_{\lambda \in \mathcal{J}}$

$$\sum_{\nu \in \mathcal{J}} |\mathbf{C}_{\lambda,\nu}| \omega_\nu \le C\,\omega_\lambda, \quad \sum_{\lambda \in \mathcal{J}} |\mathbf{C}_{\lambda,\nu}| \omega_\lambda \le C\,\omega_\nu, \quad \nu \in \mathcal{J},$$

then $\|\mathbf{C}\|_{\ell_2 \to \ell_2} \leq C$, *see e.g.* [71].

This can be proved by establishing ℓ_∞-estimates for $\mathbf{W}^{-1}\mathbf{C}\mathbf{W}$ and $\mathbf{W}^{-1}\mathbf{C}^T\mathbf{W}$, where $\mathbf{W} := \text{diag}\,(\omega_\lambda : \lambda \in \mathcal{J})$, and using then that ℓ_2 is obtained by interpolation between ℓ_1 and ℓ_∞.

In the proof of Proposition 18 this is applied to $\|\mathbf{C} - \mathbf{C}_j\|_{\ell_2}$ with weights $\omega_\lambda = 2^{-d|\lambda|/2}$ [26].

8.13 Approximation Properties and Regularity

A fundamental statement in approximation theory is that approximation properties can be expressed in terms of the *regularity* of the approximated function. In order to judge the performance of an adaptive scheme versus a much simpler scheme based on uniform refinements say, requires, in view of Theorem 8, comparing the approximation power of best N-term approximation versus approximations based on uniform refinements. The latter ones are governed essentially by *regularity* in L_2. Convergence rates N^{-s} on uniform grids with respect to some energy norm $\|\cdot\|_{H^t}$ are obtained essentially if and only if the approximated function belongs to H^{t+ds}. On the other hand, the same rate can still be achieved by best N-term approximation as long as the approximated functions belong to the (much larger) space $B_\tau^{t+sd}(L_\tau)$ where $\tau^{-1} = s + 1/2$.

The connection of these facts with the present adaptive schemes is clear from the remarks in Section 8.12. In fact, (113) says that those sequences for which $\sigma_{N,\ell_2}(\mathbf{v})$ decays like N^{-s} are (almost) those in ℓ_τ. On the other hand, ℓ_τ is directly related to regularity through relations of the type (27). Here the following version is relevant which refers to measuring the error in $\mathcal{H} = H^t$, say, see [34]. In fact, when $\mathcal{H} = H^t$, $\mathbf{D} = \mathbf{D}^t$ and $\mathbf{D}^{-t}\Psi$ is a Riesz basis for H^t, one has

$$\mathbf{u} \in \ell_\tau \quad \Longleftrightarrow \quad u = \sum_\lambda u_\lambda 2^{-t|\lambda|}\psi_\lambda \in B_\tau^{t+sd}(L_\tau(\Omega)),$$

where again $\frac{1}{\tau} = s + \frac{1}{2}$. Thus, functions in the latter space, that do not belong to H^{t+ds} can be recovered by the adaptive scheme at an *asymptotically* better rate when compared with uniform refinements. The situation is illustrated again by Figure 7 which indicates the topography of function spaces. While in Figure 4 embedding in L_2 mattered, Figure 7 shows embedding in H^t. The larger $r = t + sd$ the bigger the gap between H^r and $B_\tau^r(L_\tau)$. The loss of regularity when moving to the right from H^r at height r is compensated by judiciously placing the degrees of freedom through nonlinear approximation. Moreover, Theorem 8 says that this is preserved by the adaptive scheme.

Now the question is whether and under which circumstances the solutions to (54) have higher *Besov-regularity* than *Sobolev-regularity* to take full advantage of adaptive solvers. For scalar elliptic problems this problem has been treated e.g. in [32, 33]. The result essentially says that for rough boundaries such as Lipschitz or even polygonal boundaries the Besov-regularity of solutions is indeed higher than the relevant Sobolev-regularity, which indicates the effective use of adaptive techniques.

The analysis also shows that the quantitative performance of the adaptive scheme, compared with one based on uniform refinements, is the better the larger the H^r-norm of the solution is compared with its $B^r_\tau(L_\tau)$-norm.

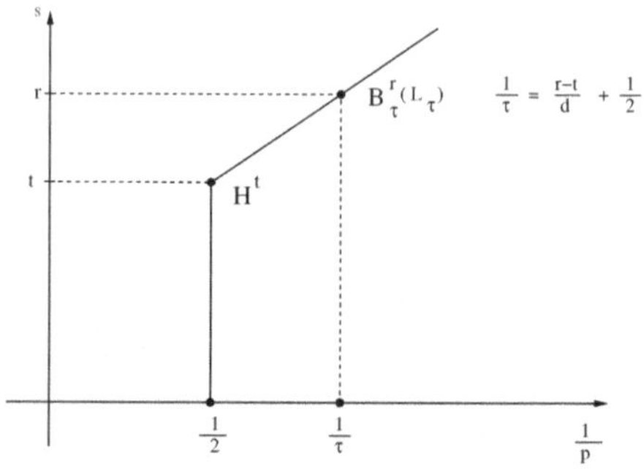

Fig. 7. Embedding in H^t

Similar results can be found for the *Stokes System* [32, 36, 33]. We know that, if the solution $U = (u, p)$ of (75) satisfies

$$u \in B^{1+sd}_\tau(L_\tau(\Omega)), \quad p \in B^{sd}_\tau(L_\tau(\Omega)), \quad \frac{1}{\tau} = s + \frac{1}{2}, \tag{118}$$

the solution components satisfy

$$\sigma_{N,H^1_0(\Omega)}(u) \lesssim N^{-s}, \quad \sigma_{N,L_2(\Omega)}(q) \lesssim N^{-s}.$$

Thus, again the question arises under which circumstances has the solution to the Stokes problem a high Besov regularity, which according to Theorem 8 and (118), would result in correspondingly high convergence rates provided by the adaptive scheme. The following result can be found in [36].

Theorem 11. *For $d = 2$ the strongest singularity solutions (u_S, p_S) of the Stokes problem on an L-shaped domain in $I\!R^2$ belong to the above scale of Besov spaces for any $s > 0$. The Sobolev regularity is limited by* 1.544483..., *resp.* 0.544483.... *Thus arbitrarily high asymptotic rates can be obtained by adaptive schemes of correspondingly high order.*

Numerical experiments for the adaptive Uzawa scheme from Section 8.4 (see (100)) for the situation described in Theorem 11 are reported in [36]. Therefore we give here only some brief excerpts. The first example concerns a strong singularity of the pressure component. Graphs of the solution components are depicted in Figure 8.

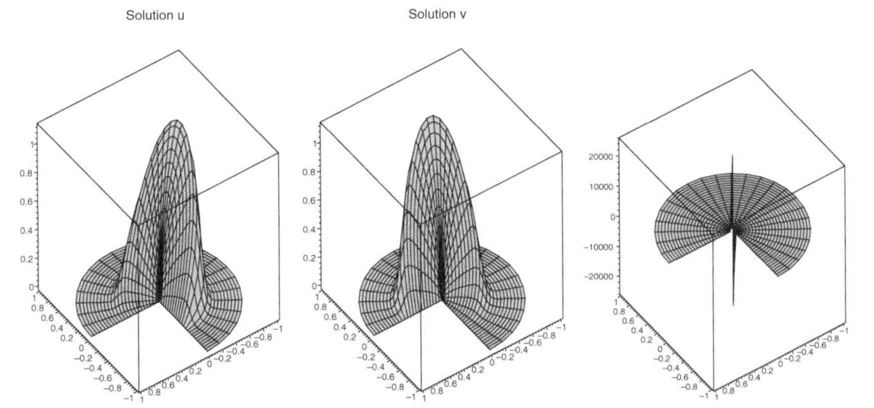

Fig. 8. Exact solution for the first example. Velocity components (left and middle) and pressure (right). The pressure functions exhibits a strong singularity

The second example involves a pressure which is *localized* around the reentrant corner, has strong gradients but is smooth, see Figure 9. To assess the

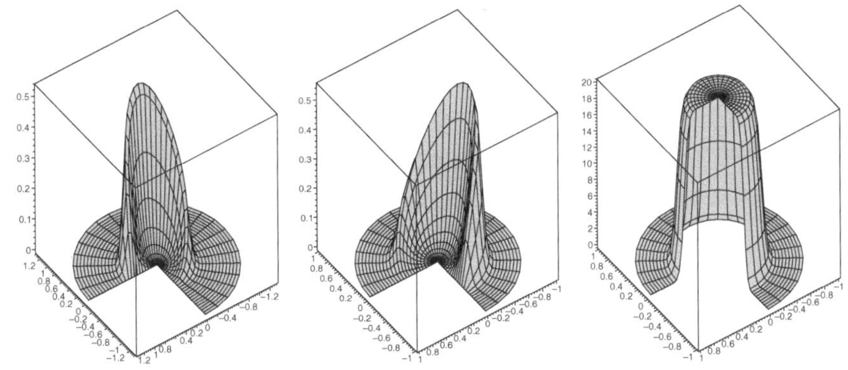

Fig. 9. Exact solution for the second example. Velocity components (left and middle) and pressure (right).

quantitative performance of the adaptive schemes, we relate the errors for the approximations produced by the scheme with the corresponding best N-term approximation error and define

$$\rho_{\mathbf{x}} := \frac{\|\mathbf{x} - \mathbf{x}_\Lambda\|_{\ell_2}}{\|\mathbf{x} - \mathbf{x}_{\#\Lambda}\|_{\ell_2}}, \qquad r_{\mathbf{x}} := \frac{\|\mathbf{x} - \mathbf{x}_\Lambda\|_{\ell_2}}{\|\mathbf{x}\|_{\ell_2}}.$$

A detailed discussion of the results for the first example are given in [36]. The factors $\rho_{\mathbf{x}}$ for velocity and pressure stay clearly below 1.5, except on the first three refinement levels. This is somewhat due to the provisional way of factoring out the constants in the pressure component which pulls in the full level of scaling functions.

Therefore we concentrate here only on an experiment for the second example, where the wavelet bases Ψ_X and Ψ_M are chosen in such a way that for fixed trial spaces the LBB-condition would be *violated.* The results are shown in Table 1, where Λ_x refers to the support of the component x. The results fully

It	$\#\Lambda_u$	$\rho_{\mathbf{u}}$	$r_{\mathbf{u}}$	$\#\Lambda_v$	$\rho_{\mathbf{v}}$	$r_{\mathbf{v}}$	$\#\Lambda_p$	$\rho_{\mathbf{p}}$	$r_{\mathbf{p}}$
3	5	1.00	0.7586	5	1.00	0.7588	243	2.23810	0.1196
4	20	1.13	0.4064	24	1.45	0.3979	262	2.08107	0.0612
5	61	1.47	0.2107	77	1.79	0.2107	324	2.72102	0.0339
6	178	1.33	0.1060	198	1.52	0.1306	396	2.81079	0.0209
7	294	1.19	0.0533	286	1.46	0.0744	674	2.21371	0.0108
8	478	1.25	0.0271	531	1.46	0.0362	899	1.83271	0.0071

Table 1. Results for the second example with piecewise linear trial functions for velocity and pressure - LBB condition is violated

confirm the theoretical predictions concerning the role of the LBB-condition. The factor ρ_p is only slightly larger than for LBB-compatible choices of Ψ_X and Ψ_M.

9 Further Issues, Applications

We shall briefly touch upon several further issues that are currently under investigation or suggest themselves from the previous developments.

9.1 Nonlinear Problems

A natural step is to apply the same paradigm to nonlinear variational problems which is done in [28, 29]. We shall briefly indicate some of the ideas for the simple example of the following boundary value problem

$$-\Delta u + u^3 = f \quad \text{in } \Omega, \quad u = 0 \quad \text{on } \Gamma = \partial\Omega. \tag{119}$$

A natural weak formulation would be to look for $u \in \mathcal{H} = H_0^1(\Omega)$ such that

$$a(v, u) := \langle \nabla v, \nabla u \rangle + \langle v, u^3 \rangle = \langle v, f \rangle, \quad v \in H_0^1(\Omega). \tag{120}$$

This presupposes, of course, that $u^3 \in H^{-1}(\Omega)$, i.e., that the mapping $u \mapsto u^3$ takes $H_0^1(\Omega)$ into $H^{-1}(\Omega)$. By simple duality arguments this can indeed be confirmed to be the case as long as the spatial dimension d is bounded by 4. It is then not hard to show that (120) is the Euler equation for the minimization of a strictly convex functional and thus turns out to possess a unique solution.

The next step is to transform (120) into wavelet coordinates

$$\boldsymbol{R}(\mathbf{u}) = \mathbf{0}, \tag{121}$$

where

$$\boldsymbol{R}(u) = \boldsymbol{A}\mathbf{u} + \mathbf{D}^{-1}\langle \Psi, u^3\rangle - \mathbf{f},$$

and with $(\mathbf{D})_{\lambda,\nu} = \delta_{\lambda,\nu}2^{|\lambda|}$

$$\boldsymbol{A} = \mathbf{D}^{-1}\langle \nabla\Psi, \nabla\Psi\rangle\mathbf{D}^{-1}, \quad \mathbf{f} = \mathbf{D}^{-1}\langle \Psi, f\rangle.$$

Now (121) is to be treated by an iterative method. The simplest version is a gradient iteration of the form

$$\mathbf{u}^{n+1} = \mathbf{u}^n - \omega\boldsymbol{R}(u^n), \quad n = 0, 1, 2, \ldots \tag{122}$$

In fact, taking the bounded invertibility of $\boldsymbol{A}$ on ℓ_2 as well as the above mentioned mapping property of the nonlinearity into account, one can show that for a suitable damping parameter ω the error is reduced in each step by a factor $\rho < 1$.

As before the objective is to carry out the iteration approximately by an adaptive application of the involved operators. The application of $\boldsymbol{A}$ can be based again on the routine **MULT** discussed above. The treatment of the nonlinearity raises several issues:

(1) Knowing the significant coefficients of $\mathbf{v} \in \ell_2$ predict the location of the significant coefficients of $\boldsymbol{R}(\mathbf{v})$, so that an overall accuracy tolerance is met.
(2) Knowing the location of the significant coefficients, compute them efficiently.

Problem (1) is treated for a class of nonlinearities in [29]. The resulting algorithm is then used in [28] to devise adaptive algorithms that exhibit again in some range asymptotically optimal work/accuracy balances, provided that the significant coefficients in the nonlinearity can be computed efficiently.

This latter task (2), in turn, can be attacked by the techniques from [48, 10]. The main idea can be sketched as follows. The searched for array

$$\mathbf{w} := \mathbf{F}(\mathbf{u}) = \mathbf{D}^{-1}\langle \Psi, F(u)\rangle, \quad F(u) := u^3,$$

consists of the *dual wavelet coefficients* of the function $w := \mathbf{w}^T\tilde{\Psi} \in L_2(\Omega)$. Thus, as soon as one has some approximation $\tilde{w}$ to w in L_2, the Riesz-basis property says that

$$\|\mathbf{w} - \tilde{\mathbf{w}}\|_{\ell_2} \lesssim \|w - \tilde{w}\|_{L_2},$$

that is, one also has a good approximation $\tilde{\mathbf{w}}$ to $\mathbf{w}$ in ℓ_2 when $\tilde{\mathbf{w}}$ are the dual wavelet coefficients of the approximation $\tilde{w}$. The objective is now to construct

$\tilde{w}$ with the aid of suitable quadrature techniques and quasi-interpolants in combination with *local single scale representations.* The array $\tilde{\mathbf{w}}$ is then obtained by local multiscale transformations. An important point is that quadrature is not used for the approximation of the individual entries but for determining the global approximant $\tilde{w}$. For details the reader is referred to [48, 10].

9.2 Time Dependent Problems

The simplest example of a time dependent problem, that can be attacked by the above concepts, is the heat equation or more generally

$$\partial_t u = \mathcal{L}u,$$

where $\mathcal{L}$ is an elliptic operator (incorporating as above homogeneous boundary conditions). Using an implicit time discretization, the method of lines turns this into a series of elliptic boundary value problems, which can be treated by the concepts discussed above.

An alternative is to write

$$u(t) = e^{t\mathcal{L}}u(0)$$

and utilize the following representation of the exponential

$$e^{t\mathcal{L}}u_0 = \frac{1}{2\pi i}\int_{\Gamma} e^{tz}(zI - \mathcal{L})^{-1}u_0 dz,$$

which in a somewhat different context has been also done in [58]. Here Γ is a curve in the complex plane avoiding the spectrum of $\mathcal{L}$. Approximating the right hand side by quadrature, we obtain

$$u(t) \approx \sum_n \omega_n e^{tz_n}(z_n\mathbf{I} - \mathcal{L})^{-1}u_0,$$

see [58, 63]. Thus one has can solve the individual problems $(z_n\mathbf{I} - \mathcal{L})u_{(n)} = u_0$ in parallel.

10 Appendix: Some Useful Facts

For the convenience of the reader a few useful facts are summarized in this section which have been frequently referred to in previous sections.

10.1 Function Spaces

We collect first a few definitions concerning function spaces. Standard references are [1, 11, 73].

Sobolev spaces of integer order k (on bounded domains or $\mathbb{R}^d$) are defined as $W_p^k(\Omega) := \{f : \partial^\alpha f \in L_p(\Omega),\ |\alpha| \leq k\}$, where derivatives are understood in the weak sense. The corresponding (semi-)norms are given by $|f|_{W_p^k(\Omega)} := \left(\sum_{|\alpha|=k} \|\partial^\alpha f\|^p_{L_p(\Omega)}\right)^{1/p}$, $\|v\|^p_{W_p^k(\Omega)} := \sum_{m=0}^k |f|^p_{W_p^m(\Omega)}$.

Fractional order spaces can be defined by *intrinsic norms* with $k := \lfloor t \rfloor$:

$$\|v\|_{W_p^t(\Omega)} = \left(\|v\|^p_{W_p^k(\Omega)} + \sum_{|\alpha|=k} \int_\Omega \int_\Omega \frac{|\partial^\alpha v(x) - \partial^\alpha v(y)|^p}{|x-y|^{d+tp}} dxdy \right)^{1/p}.$$

Negative indices are handled by duality.

An alternative is offered by extension to $\mathbb{R}^d$ and using then Fourier transforms or by *interpolation*. This leads to the notion of *Besov-spaces*. For a given fixed r and any $t < r$ the quantities

$$|v|_{B_q^t(L_p(\Omega))} := \begin{cases} \left(\int\limits_0^\infty [s^{-t}\omega_r(f,s,\Omega)_p]^q \frac{ds}{s} \right)^{1/q}, & 0 < q < \infty; \\ \sup_{s>0} s^{-t}\omega_r(f,s,\Omega)_p, & q = \infty, \end{cases}$$

define equivalent Besov semi-norms. Here we have employed the *L_p-modulus* of continuity:

$$\omega_r(f,t,\Omega)_p := \sup_{|h|\leq t} \|\Delta_h^r f\|_{L_p(\Omega_{r,h})},$$

where

$$\Delta_h f := f(\cdot + h) - f(\cdot), \quad \Delta_h^k = \Delta_h \circ \Delta_h^{k-1}$$

and $\Omega_{r,h} := \{x : x + sh \in \Omega,\ s \in [0,r]\}$.

Important *special cases* are:

- $W_p^t = B_p^t(L_p)$ for $t > 0$, $s \notin \mathbb{N}$, $(p \neq 2)$
- $H^t := W_2^t = B_2^t(L_2)$ for $t \in \mathbb{R}$ $\quad (H^{-t} := (H^t)')$, where as before

$$\|f\|_{X'} := \sup_{g\in X, g\neq 0} \frac{\langle f,g\rangle}{\|g\|_X}.$$

10.2 Local Polynomial Approximation

Estimates of the following type are frequently used, see e.g. [55].

$$\begin{aligned} \inf_{P\in\mathbb{P}_k} \|v-P\|_{W_p^m(\Omega)} &\lesssim (\operatorname{diam}\Omega)^{k-m}|v|_{W_p^k(\Omega)}, \quad (m<k) \\ \inf_{P\in\mathbb{P}_k} \|v-P\|_{L_p(\Omega)} &\lesssim (\operatorname{diam}\Omega)^t|v|_{B_q^t(L_p(\Omega))}, \\ \inf_{P\in\mathbb{P}_k} \|v-P\|_{L_p(\Omega)} &\lesssim \sup_{t>0}\omega_k(f,t,\Omega)_p. \end{aligned}$$

Idea of proof: By rescaling it suffices to consider a reference domain with unit diameter. Suppose that $\inf_{P\in\mathbb{P}_k}\|v_n - P\|_{W_p^m(\Omega)} \geq n|v_n|_{W_p^k(\Omega)}$. Rescale to conclude that

$$1 = \inf_{P\in\mathbb{P}_k}\|w_n - P\|_{W_p^m(\Omega)} = \|w_n\|_{W_p^m(\Omega)} \geq n|w_n|_{W_p^k(\Omega)}.$$

Thus $\{w_n\}_n$ is precompact in W_p^m, that is there exists $w \in W_p^m$ such that

$$|w|_{W_p^k(\Omega)} = \lim_{n\to\infty}|w_n|_{W_p^k(\Omega)} = 0,$$

which implies that $w \in \mathbb{P}_k$. On the other hand, $\inf_{P\in\mathbb{P}_k}\|w - P\|_{W_p^m(\Omega)} = 1$ which is a contradiction. This provides the first estimate. The remaining cases are similar. □

10.3 Condition Numbers

The connection between the order of an operator and the condition numbers of related stiffness matrices can be described as follows.

Remark 19 *If* $a(\cdot,\cdot) \sim \|\cdot\|_{H^t}^2$ *is symmetric and* Ψ *is an* L_2*-Riesz-basis, then for* $J := \max\{|\lambda| : \lambda \in \Lambda\}$, $\min\{|\lambda| : \lambda\in\Lambda\} \lesssim 1$, *one has*

$$\mathrm{cond}_2\left(a(\Psi_\Lambda,\Psi_\Lambda)\right) \sim 2^{2|t|J}, \quad J\to\infty.$$

Proof:

$$\max_{v\in S_\Lambda}\frac{a(v,v)}{\|v\|_{L_2}^2} \geq \frac{a(\psi_\lambda,\psi_\lambda)}{\|\psi_\lambda\|_{L_2}^2} \geq \min_{v\in S_\Lambda}\frac{a(v,v)}{\|v\|_{L_2}^2},$$

which gives

$$\mathrm{cond}_2(a(\Psi_\Lambda,\Psi_\Lambda)) \gtrsim \frac{\frac{a(\psi_{\lambda_1},\psi_{\lambda_1})}{\|\psi_{\lambda_1}\|_{L_2}^2}}{\frac{a(\psi_{\lambda_2},\psi_{\lambda_2})}{\|\psi_{\lambda_2}\|_{L_2}^2}}.$$

Since $a(\psi_\lambda,\psi_\lambda) \sim \|\psi_\lambda\|_{H^t}^2$ the norm equivalences (16) imply $2^{t|\lambda|} \sim \|\psi_\lambda\|_{H^t}$. Now setting

$$|\lambda_1| = \begin{cases}\max\{|\lambda| : \lambda\in\Lambda\} \text{ if } t\geq 0,\\ \min\{|\lambda| : \lambda\in\Lambda\} \text{ if } t<0,\end{cases}$$

and

$$|\lambda_2| = \begin{cases}\min\{|\lambda| : \lambda\in\Lambda\} \text{ if } t\geq 0,\\ \max\{|\lambda| : \lambda\in\Lambda\} \text{ if } t<0,\end{cases}$$

we conclude $\mathrm{cond}_2(a(\Psi_\Lambda,\Psi_\Lambda)) \gtrsim 2^{2J|t|}$. As for the *upper estimate*, consider first the case $t<0$:

$$\min_{v\in S_\Lambda}\frac{a(v,v)}{\|v\|_{L_2}^2} \sim \min_{\boldsymbol{d}_\Lambda\in\mathbb{R}^{\#\Lambda}}\frac{\|\boldsymbol{d}_\Lambda^T\Psi_\Lambda\|_{H^t}^2}{\|\boldsymbol{d}_\Lambda\|_{\ell_2}^2} \overset{\text{(NE)}}{\sim} \min_{\boldsymbol{d}_\Lambda\in\mathbb{R}^{\#\Lambda}}\frac{\|\mathbf{D}^{-t}\boldsymbol{d}_\Lambda\|_{\ell_2}^2}{\|\boldsymbol{d}_\Lambda\|_{\ell_2}^2} \gtrsim 2^{-|t|J},$$

while

$$\max_{v\in S_\Lambda} \frac{a(v,v)}{\|v\|_{L_2}^2} \lesssim 1.$$

In the case $t > 0$, the Bernstein estimate (29) yields

$$\max_{v\in S_\Lambda} \frac{a(v,v)}{\|v\|_{L_2}^2} \lesssim \max_{v\in S_\Lambda} \frac{\|v\|_{H^t}^2}{\|v\|_{L_2}^2}, \lesssim 2^{Jt},$$

which shows that $\mathrm{cond}_2(a(\Psi_\Lambda, \Psi_\Lambda)) \lesssim 2^{2J|t|}$. □

References

1. R.A. Adams, *Sobolev Spaces*, Academic Press, 1978.
2. A. Averbuch, G. Beylkin, R. Coifman, and M. Israeli, Multiscale inversion of elliptic operators, in: Signal and Image Representation in Combined Spaces, J. Zeevi, R. Coifman (eds.), Academic Press, 1995,
3. I. Babuška, *The finite element method with Lagrange multipliers*, Numer. Math. 20, 1973, 179–192.
4. I. Babuška and W.C. Rheinboldt, Error estimates for adaptive finite element computations, SIAM J. Numer. Anal. 15 (1978), 736–754.
5. R.E. Bank and A. Weiser, Some a posteriori error estimates for elliptic partial differential equations, Math. Comp., 44 (1985), 283–301.
6. H.J.C. Barbosa and T.J.R. Hughes, Boundary Lagrange multipliers in finite element methods: Error analysis in natural norms, Numer. Math., 62 (1992), 1–15.
7. A. Barinka, Fast evaluation tools for adaptive wavelet schemes, PhD Thesis, RWTH Aachen, 2003.
8. A. Barinka, T. Barsch, P. Charton, A. Cohen, S. Dahlke, W. Dahmen, K. Urban, Adaptive wavelet schemes for elliptic problems – Implementation and numerical experiments, SIAM J. Sci. Comp., 23, No. 3 (2001), 910–939.
9. A. Barinka, S. Dahlke, W. Dahmen, Adaptive Application of Operators in Standard Representation, in preparation.
10. A. Barinka, W. Dahmen, R. Schneider, Y. Xu, An algorithm for the evaluation of nonlinear functionals of wavelet expansions, in preparation
11. J. Bergh, J. Löfström, *Interpolation Spaces, An Introduction*, Springer, 1976.
12. S. Bertoluzza, A-posteriori error estimates for wavelet Galerkin methods, Appl. Math. Lett., 8 (1995), 1–6.
13. G. Beylkin, R. R. Coifman, V. Rokhlin, Fast wavelet transforms and numerical algorithms I, Comm. Pure and Appl. Math., 44 (1991), 141–183.
14. G. Beylkin, J. M. Keiser, An adaptive pseudo-wavelet approach for solving nonlinear partial differential equations, in: *Multiscale Wavelet Methods for PDEs*, W. Dahmen, A. J. Kurdila, P. Oswald (eds.) Academic Press, 137–197, 1997.
15. F. Bornemann, B. Erdmann, and R. Kornhuber, A posteriori error estimates for elliptic problems in two and three space dimensions, SIAM J. Numer. Anal., 33 (1996), 1188–1204.
16. J.H. Bramble, The Lagrange multiplier method for Dirichlet's problem, Math. Comp., 37 (1981), 1–11.

17. J.H. Bramble, R.D. Lazarov, and J.E. Pasciak, A least-squares approach based on a discrete minus one inner product for first order systems, Math. Comput., 66 (1997), 935–955.
18. J.H. Bramble, R.D. Lazarov, and J.E. Pasciak, Least-squares for second order elliptic problems, Comp. Meth. Appl. Mech. Engnrg., 152 (1998), 195–210.
19. F. Brezzi and M. Fortin, *Mixed and Hybrid Finite Element Methods*, Springer, 1991.
20. Z. Cai, R. Lazarov, T.A. Manteuffel, and S.F. McCormick, First-order system least squares for second-order partial differential equations; Part I, SIAM J. Numer. Anal., 31 (1994), 1785–1799.
21. A. Canuto, A. Tabacco, K. Urban, The wavelet element method, part I: Construction and analysis, Appl. Comp. Harm. Anal., 6 (1999), 1–52.
22. A. Canuto, A. Tabacco, K. Urban, The wavelet element method, part II: Realization and additional features in 2D and 3D, Appl. Comp. Harm. Anal., 8 (2000), 123–165.
23. J.M. Carnicer, W. Dahmen and J.M. Peña, Local decomposition of refinable spaces, Appl. Comp. Harm. Anal., 3 (1996), 127-153.
24. A. Cohen, *Wavelet methods in numerical analysis*, in the Handbook of Numerical Analysis, vol. VII, P.-G. Ciarlet et J.-L. Lions eds., Elsevier, Amsterdam, 2000.
25. A. Cohen, I. Daubechies, J.-C. Feauveau, Biorthogonal bases of compactly supported wavelets, Comm. Pure and Appl. Math., 45 (1992), 485–560.
26. A. Cohen, W. Dahmen, R. DeVore, Adaptive wavelet methods for elliptic operator equations – Convergence rates, Math. Comp., 70 (2001), 27–75.
27. A. Cohen, W. Dahmen, R. DeVore, Adaptive wavelet methods II - Beyond the elliptic case, Foundations of Computational Mathematics, 2 (2002), 203–245.
28. A. Cohen, W. Dahmen, R. DeVore, Adaptive Wavelet Schemes for Nonlinear Variational Problems, IGPM Report # 221, RWTH Aachen, July 2002.
29. A. Cohen, W. Dahmen, R. DeVore, Sparse evaluation of nonlinear functionals of multiscale expansions, IGPM Report # 222, RWTH Aachen, July 2002.
30. A. Cohen, R. Masson, Wavelet adaptive methods for second order elliptic problems: boundary conditions and domain decomposition, Numer. Math., 86 (2000), 193–238,
31. M. Costabel and E.P. Stephan, Coupling of finite and boundary element methods for an elastoplastic interface problem, SIAM J. Numer. Anal., 27 (1990), 1212–1226.
32. S. Dahlke: Besov regularity for elliptic boundary value problems on polygonal domains, Appl. Math. Lett., 12 (1999), 31–36.
33. S. Dahlke, R. DeVore, Besov regularity for elliptic boundary value problems, Comm. Partial Differential Equations, 22 (1997), 1–16.
34. S. Dahlke, W. Dahmen, R. DeVore, Nonlinear approximation and adaptive techniques for solving elliptic operator equations, in: *Multiscale Wavelet Methods for PDEs*, W. Dahmen, A. Kurdila, P. Oswald (eds.), Academic Press, London, 237–283, 1997.
35. S. Dahlke, W. Dahmen, R. Hochmuth, R. Schneider, *Stable multiscale bases and local error estimation for elliptic problems*, Appl. Numer. Maths. 8 (1997), 21–47.
36. S. Dahlke, W. Dahmen, K. Urban, Adaptive wavelet methods for saddle point problems – Convergence rates, SIAM J. Numer. Anal., 40 (No. 4) (2002), 1230–1262.

37. S. Dahlke, R. Hochmuth, K. Urban, Adaptive wavelet methods for saddle point problems, Math. Model. Numer. Anal. (M2AN), 34(5) (2000), 1003-1022.
38. W. Dahmen, Some remarks on multiscale transformations, stability and biorthogonality, in: *Wavelets, Images and Surface Fitting*, P.J. Laurent, A. Le M´ehaut´e, L.L. Schumaker (eds.), AK Peters, Wellesley, Massachusetts, 157-188, 1994.
39. W. Dahmen, Stability of multiscale transformations, Journal of Fourier Analysis and Applications, 2 (1996), 341-361.
40. W. Dahmen, Wavelet and Multiscale Methods for Operator Equations, Acta Numerica, Cambridge University Press, 6 (1997), 55–228.
41. W. Dahmen, Wavelet methods for PDEs – Some recent developments, J. Comp. Appl. Math., 128 (2001), 133–185.
42. W. Dahmen, A. Kunoth, Multilevel preconditioning, Numer. Math., 63 (1992), 315–344.
43. W. Dahmen, A. Kunoth, Appending boundary conditions by Lagrange multipliers: Analysis of the LBB condition, Numer. Math., 88 (2001), 9–42.
44. W. Dahmen, A. Kunoth, R. Schneider, Wavelet least squares methods for boundary value problems, SIAM J. Numer. Anal. 39(6) (2002), 1985–2013.
45. W. Dahmen, S. Prößdorf, R. Schneider, Multiscale methods for pseudo-differential equations on smooth manifolds, in: *Proceedings of the International Conference on Wavelets: Theory, Algorithms, and Applications*, C.K. Chui, L. Montefusco, L. Puccio (eds.), Academic Press, 385-424, 1994.
46. W. Dahmen and R. Schneider, Composite Wavelet Bases for Operator Equations, Math. Comp., 68 (1999), 1533–1567.
47. W. Dahmen and R. Schneider, Wavelets on Manifolds I: Construction and Domain Decomposition, SIAM J. Math. Anal., 31 (1999), 184–230.
48. W. Dahmen, R. Schneider, Y. Xu, Nonlinear functions of wavelet expansions – Adaptive reconstruction and fast evaluation, Numerische Mathematik, 86 (2000), 49–101.
49. W. Dahmen and R. Stevenson, Element-by-element construction of wavelets – stability and moment conditions, SIAM J. Numer. Anal., 37 (1999), 319–325.
50. I. Daubechies, *Ten Lectures on Wavelets*, CBMS-NSF Regional Conference Series in Applied Math. 61, SIAM, Philadelphia, 1992.
51. R. DeVore, Nonlinear approximation, Acta Numerica, Cambridge University Press, 7 (1998), 51-150.
52. R. DeVore, V. Popov, Interpolation of Besov spaces, Trans. Amer. Math. Soc., 305 (1988), 397–414.
53. R. DeVore, B. Jawerth, and V. Popov, Compression of wavelet decompositions, Amer. J. Math., 114 (1992), 737–785.
54. R. DeVore, G.G. Lorentz, *Constructive Approximation*, Grundlehren vol. 303, Springer-Verlag, Berlin, 1993.
55. R. DeVore, R. Sharpley, Maximal functions measuring smoothness, Memoirs of the American Mathematical Society, 47 (No 293), 1984.
56. W. Dörfler, A convergent adaptive algorithm for Poisson's equation, SIAM J. Numer. Anal., 33 (1996), 1106–1124.
57. K. Eriksson, D. Estep, P. Hansbo, and C. Johnson, Introduction to adaptive methods for differential equations, Acta Numerica, Cambridge University Press, 4 (1995), 105–158.

58. I.P. Gavrilyuk, W. Hackbusch, B.N. Khoromskij, H-matrix approximation for the operator exponential with applications, Preprint No. 42, Max-Planck-Institut für Mathematik, Leipzig, 2000, to appear in Numer. Math..
59. V. Girault and P.-A. Raviart, *Finite Element Methods for Navier-Stokes Equations*, Springer, 1986.
60. R. Glowinski and V. Girault, *Error analysis of a fictitious domain method applied to a Dirichlet problem*, Japan J. Industr. Appl. Maths. 12 (1995), 487–514.
61. M.D. Gunzburger, S.L. Hou, Treating inhomogeneous boundary conditions in finite element methods and the calculation of boundary stresses, SIAM J. Numer. Anal., 29 (1992), 390–424.
62. H. Harbrecht, *Wavelet Galerkin Schemes for the Boundary Element Method in three Dimensions*, Doctoral Dissertation, Technical University of Chemnitz, June 2001.
63. M. Jürgens, Adaptive methods for time dependent problems, in preparation.
64. R. Kress, Linear Integral Equations, Springer-Verlag, Berlin-Heidelberg, 1989.
65. E. Novak, On the power of adaptation, J. Complexity, 12 (1996), 199–237.
66. T. von Petersdorff, C. Schwab, Wavelet approximation for first kind integral equations on polygons, Numer. Math., 74 (1996) 479-516.
67. T. von Petersdorff, C. Schwab, Fully discrete multiscale Galerkin BEM, in: *Multiscale Wavelet Methods for PDEs*, W. Dahmen, A. Kurdila, P. Oswald (eds.), Academic Press, London, 287–346, 1997.
68. R. Schneider, *Multiskalen- und Wavelet-Matrixkompression: Analysisbasierte Methoden zur effizienten Lösung großer vollbesetzter Gleichungssysteme*, Habilitationsschrift, Technische Hochschule, Darmstadt, 1995, *Advances in Numerical Mathematics*, Teubner,1998.
69. W. Sweldens, The lifting scheme: A custom-design construction of biorthogonal wavelets, Appl. Comput. Harm. Anal., 3 (1996), 186-200.
70. W. Sweldens, The lifting scheme: A construction of second generation wavelets, SIAM J. Math. Anal., 29 (1998), 511–546.
71. P. Tchamitchian, Wavelets, Functions, and Operators, in: *Wavelets: Theory and Applications*, G. Erlebacher, M.Y. Hussaini, and L. Jameson (eds.), ICASE/LaRC Series in Computational Science and Engineering, Oxford University Press, 83–181, 1996.
72. J. Traub and H. Woźniakowski, *A General Theory of Optimal Algorithms*, Academic Press, New York, 1980.
73. H. Triebel, *Interpolation Theory, Function Spaces, and Differential Operators*, North-Holland, Amsterdam, 1978.
74. P.S. Vassilevski, J. Wang, Stabilizing the hierarchical basis by approximate wavelets, I: Theory, Numer. Linear Algebra Appl., 4 (1997),103-126.
75. H. Yserentant, On the multilevel splitting of finite element spaces, Numer. Math. 49 (1986), 379–412.

Multilevel Methods in Finite Elements

James H. Bramble

Mathematics Department, Texas A&M University, College Station, Texas, USA
bramble@math.tamu.edu

Summary. This survey consists of five parts. In the first section we describe a model problem and a two-level algorithm in order to motivate the multilevel approach. In Section 2 an abstract multilevel algorithm is described and analyzed under some regularity assumptions. An analysis under less stringent assumptions is given in Section 3. Non-nested spaces and varying forms are treated in Section 4. Finally, we show how the multilevel framework provides computationally efficient realizations of norms on Sobolev scales.

1 Introduction

We consider a two-level multigrid algorithm applied to a simple model problem in this section. The purpose here is to describe and motivate the use of multiple grids. We will give a complete analysis of this algorithm. For this purpose and for the treatment of multilevel methods, we first provide some preliminary definitions. Next, a model problem and its finite element approximation are described. The two-level method is then defined and an analysis of its properties as a reducer/preconditioner is provided.

1.1 Sobolev Spaces

The iterative convergence estimates for multigrid algorithms applied to the computation of the discrete approximations to partial differential equations are most naturally analyzed using Sobolev spaces and their associated norms. To be precise, we shall give the definitions here although a more thorough discussion can be found in, for example, [1], [14] and [19].

Let Ω be a Lebesgue measurable set in d dimensional Euclidean space $\mathbf{R}^d$ and f be a real valued function defined on Ω. We denote by

$$\|f\|_{L_p(\Omega)} = \left(\int_{\Omega} |f(x)|^p \, \mathrm{d}x \right)^{1/p}$$

the $L_p(\Omega)$ norm of f. Let $\mathbf{N}$ denote the set of nonnegative integers and let $\alpha = (\alpha_1, \ldots, \alpha_N)$, with $\alpha_i \in \mathbf{N}$, be a multi-index. We set

$$|\alpha| = \alpha_1 + \cdots + \alpha_N$$

and

$$D^\alpha = \left(\frac{\partial}{\partial x_1}\right)^{\alpha_1} \left(\frac{\partial}{\partial x_2}\right)^{\alpha_2} \cdots \left(\frac{\partial}{\partial x_N}\right)^{\alpha_N}.$$

The Sobolev spaces $W_p^s(\Omega)$ for $s \in \mathbf{N}$ are defined to be the set of distributions $f \in \mathcal{D}'(\Omega)$ (cf. [17]) for which the norm

$$\|f\|_{W_p^s(\Omega)} = \left(\sum_{|\alpha| \leq s} \|D^\alpha f\|_{L_p(\Omega)}^p \right)^{1/p} < \infty.$$

When $p = 2$, the spaces are Hilbert spaces and are of special interest in this book. We shall denote these by $H^s(\Omega) \equiv W_2^s(\Omega)$. The corresponding norm will be denoted by

$$\|\cdot\|_{s,\Omega} = \|\cdot\|_{W_2^s(\Omega)}.$$

For real s with $i < s < i+1$, the Sobolev space $H^s(\Omega)$ is defined by interpolation (using the real method) between $H^i(\Omega)$ and $H^{i+1}(\Omega)$ (see, e.g., [12], [19] or Appendix A of [3]). The norm and inner product notation will be further simplified when the domain Ω is clear from the context, in which case we use

$$\|\cdot\|_s = \|\cdot\|_{s,\Omega} \quad \text{and} \quad (\cdot,\cdot) = (\cdot,\cdot)_\Omega.$$

We will also use certain Sobolev spaces with negative indices. These will be defined later as needed.

1.2 A Model Problem

In this section we consider the Dirichlet problem on a bounded domain Ω in $\mathbf{R}^2$. This problem and its finite element approximation can be used to illustrate some of the most fundamental properties of the multigrid algorithms. Let

$$\Delta = \frac{\partial^2}{\partial x^2} + \frac{\partial^2}{\partial y^2}.$$

Given f in an appropriately defined Sobolev space, consider the Dirichlet problem

$$\begin{cases} -\Delta u = f, & \text{in } \Omega, \\ \quad u = 0, & \text{on } \partial\Omega. \end{cases} \tag{1}$$

For $v, w \in H^1(\Omega)$, let

$$D(v,w) = \int_\Omega \left(\frac{\partial v}{\partial x}\frac{\partial w}{\partial x} + \frac{\partial v}{\partial y}\frac{\partial w}{\partial y} \right) dx\,dy.$$

Denote by $C_0^\infty(\Omega)$ the space of infinitely differentiable functions with compact support in Ω. By Green's identity, for $\phi \in C_0^\infty(\Omega)$,

$$(f, \phi) = (-\Delta u, \phi) = D(u, \phi). \tag{2}$$

Let $H_0^1(\Omega)$ be the closure of $C_0^\infty(\Omega)$ with respect to $\|\cdot\|_1$. The Poincaré inequality implies that there is a constant $C > 0$ such that

$$\|v\|_0^2 \le CD(v, v), \qquad \text{for all } v \in H_0^1(\Omega).$$

Hence, we can take $D(\cdot, \cdot)^{1/2}$ to be the norm on $H_0^1(\Omega)$. This changes the Hilbert space structure.

In the above inequality, C represents a generic positive constant. Such constants will appear often in this book and will be denoted by C and c, with or without subscript. These constants can take on different values at different occurrences, however, they will always be independent of mesh and grid level parameters.

For a bounded linear functional f on $H_0^1(\Omega)$, the weak solution u of (1) satisfies

$$D(u, \phi) = (f, \phi), \qquad \text{for all } \phi \in H_0^1(\Omega). \tag{3}$$

Here (f, ϕ) is the value of the functional f at ϕ. If $f \in L^2(\Omega)$, it coincides with the L^2-inner product. Existence and uniqueness of the function $u \in H_0^1(\Omega)$ satisfying (3) will follow from the Poincaré inequality and the Riesz Representation Theorem.

Theorem 1.1 (Riesz Representation Theorem) *Let H be a Hilbert space with norm $\|\cdot\|$ and inner product $(\cdot, \cdot)_H$. Let f be a bounded linear functional on H, i.e.,*

$$|f(\phi)| \le C(f)\|\phi\|.$$

Then there exists a unique $u_f \in H$ such that

$$(u_f, \phi)_H = f(\phi), \qquad \textit{for all } \phi \in H.$$

To apply the above theorem to (3), we take $H = H_0^1(\Omega)$ with $(\cdot, \cdot)_H = D(\cdot, \cdot)$ and set

$$f(\phi) = (f, \phi), \qquad \text{for all } \phi \in H_0^1(\Omega).$$

Then, by the definition of $\|\cdot\|_{(H_0^1(\Omega))'}$ and the Poincaré inequality,

$$|f(\phi)| \le \|f\|_{(H_0^1(\Omega))'} \|\phi\|_1 \le C\|f\|_{(H_0^1(\Omega))'} D(\phi, \phi)^{1/2}.$$

Thus, $f(\cdot)$ is a bounded linear functional on H and Theorem 1.1 implies that there is a unique function $u \in H = H_0^1(\Omega)$ satisfying (3).

1.3 Finite Element Approximation of the Model Problem

In this subsection, we consider the simplest multilevel finite element approximation spaces. We start with the Galerkin Method. subspace of $H_0^1(\Omega)$. The Galerkin approximation is the function $u \in M$ satisfying

$$D(u, \phi) = (f, \phi), \qquad \text{for all } \phi \in M. \tag{4}$$

As in the continuous case, the existence and uniqueness of solutions to (4) follows from the Poincaré inequality and the Riesz Representation Theorem. We shall consider spaces M which result from multilevel finite element constructions.

We first define a nested sequence of triangulations. Let Ω be a domain with polygonal boundary and let $\mathcal{T}_1$ be a given (coarse) triangulation of Ω. Successively finer triangulations $\{\mathcal{T}_k\}$, $k = 2, \ldots, J$ are formed by subdividing the triangles of $\mathcal{T}_{k-1}$. More precisely, for each triangle τ of $\mathcal{T}_k$, $\mathcal{T}_{k+1}$ has four triangles corresponding to those formed by connecting the midpoints of the sides of τ. Note that the angles of the triangles in the finest triangulation are the same as those in the coarsest. Thus, the triangles on all grids are of quasi-uniform shape independent of the mesh parameter k. This construction is illustrated in Figure 1. Nested approximation spaces are defined in terms

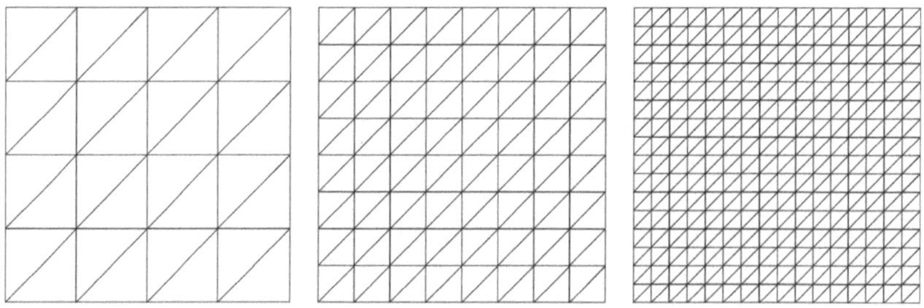

Fig. 1. Nested triangulations

of the triangulations. Let M_k, for $k = 1, \ldots, J$, be the space of continuous piecewise linear functions on $\mathcal{T}_k$ which vanish on $\partial\Omega$. Set h_1 to be the length of the side of maximum length in $\mathcal{T}_1$ and $h_k = 2^{-k+1}h_1$. We clearly have that

$$M_1 \subset M_2 \subset \cdots \subset M_J \equiv M \subset H_0^1(\Omega).$$

We also denote $h = h_J$.

The sequence of spaces defined above satisfy the following approximation properties: For $v \in H_0^1(\Omega) \cap H^r(\Omega)$ with $r = 1$ or 2, there exists $\chi \in M_k$ such that

$$\|v - \chi\|_0^2 + h_k^2 \|v - \chi\|_1^2 \leq C h_k^{2r} \|v\|_r^2. \tag{5}$$

See, e.g., [11] for a proof.

1.4 The Stiffness Matrix and its Condition Number

The point of our multigrid algorithms is to develop effective iterative techniques for computing approximate solutions satisfying equations exemplified by (4). Functions in M are most naturally represented in terms of a basis. Order the interior vertices of $\mathcal{T}_J$, $1 \leq i \ldots \leq N_J$ and let $\phi_i \in M$ be such that

$$\phi_i = \begin{cases} 1 & \text{at } x_i \\ 0 & \text{at } x_j \neq x_i. \end{cases}$$

Since any continuous, piecewise linear function is determined by its values at the vertices,

$$u = \sum_{i=1}^{N_J} \tilde{u}_i \, \phi_i \quad \text{where} \quad \tilde{u}_i = u(x_i).$$

Hence,

$$\sum_{i=1}^{N_J} \tilde{u}_i \, D(\phi_i, \phi_j) = (f, \phi_j). \tag{6}$$

Let $\underset{\approx}{A}_J$ denote the stiffness matrix $[\underset{\approx}{A}_J]_{ij} = D(\phi_i, \phi_j)$. Equation (6) is equivalent to

$$\underset{\approx}{A}_J \tilde{u} = \underset{\sim}{f}, \tag{7}$$

where $\tilde{u} = (\tilde{u}_1, \ldots, \tilde{u}_N)^T$, $\underset{\sim}{f} = (\underset{\sim}{f}_1, \ldots, \underset{\sim}{f}_N)^T$ and $\underset{\sim}{f}_i = (f, \phi_i)$.

It is well known that the rate of convergence of simple linear iterative methods or the conjugate gradient method, directly applied to (7), can be bounded in terms of the condition number of $\underset{\approx}{A}_J$ (cf. [3]). We will now estimate this condition number.

Let v be in M_J and write

$$v = \sum_{i=1}^{N_J} \tilde{v}_i \, \phi_i.$$

Let τ be a triangle of $\mathcal{T}_J$. Since τ is of quasi-uniform shape and v is linear on τ, we clearly have that $\|v\|_{0,\tau}^2$ is equivalent to the sum of the squares of the values of v at the vertices of τ times h^2. By summing over the triangles,

$$\|v\|_0^2 \approx h^2 \sum_{i=1}^{N_J} \tilde{v}_i^2, \qquad v \in M_J.$$

The notation $\approx$ means equivalence of norms with constants of equivalence independent of h. Hence, by the Poincaré inequality,

$$h^2 \sum_{i=1}^{N_J} \tilde{v}_i^2 \leq C \|v\|_0^2 \leq C D(v, v) = C \sum_{i,j=1}^{N_J} [\underset{\approx}{A}_J]_{ij} \, \tilde{v}_i \tilde{v}_j, \qquad v \in M_J.$$

This means that the smallest eigenvalue of $\underset{\approx}{A}_J$ is bounded below by ch^2.

Let τ be a triangle in $\mathcal{T}_J$. Since v is linear on τ and τ is of quasi-uniform shape, we have

$$\int_\tau |\nabla v|^2 \, dx \approx [v(a) - v(b)]^2 + [v(b) - v(c)]^2, \qquad v \in M_J.$$

Here a, b, c are the vertices of τ. By summing and using obvious manipulations, it follows that

$$D(v, v) \leq C_1 \sum_{i=1}^{N_J} \tilde{v}_i^2, \qquad \text{for all } v \in M_J.$$

This means that the largest eigenvalue is bounded and hence the condition number of $\underset{\approx}{A}_J$ is bounded by

$$\kappa(\underset{\approx}{A}_J) \leq Ch^{-2}. \tag{8}$$

It is not difficult to show that the estimate (8) is sharp, i.e., there is a constant c not depending on J such that

$$\kappa(\underset{\approx}{A}_J) \geq ch^{-2}.$$

We leave this as an exercise for the reader. This means that the problem is rather ill conditioned and, thus, iterative methods applied directly to (7) will, in general, converge slowly.

1.5 A Two-Level Multigrid Method

To motivate the two-level multigrid method, we start by considering the simple linear iterative method

$$\tilde{u}^{n+1} = \tilde{u}^n - \underset{\approx}{\lambda}_J^{-1}(\underset{\approx}{A}_J \tilde{u}^n - \underset{\sim}{f}). \tag{9}$$

Here $\underset{\approx}{\lambda}_J$ denotes the largest eigenvalue of $\underset{\approx}{A}_J$. Since $\underset{\approx}{A}_J$ is symmetric and positive definite, there is a basis $\{\tilde{\psi}_i\}$, for $i = 1, \dots, N$ of eigenfunctions with corresponding eigenvalues $0 < \eta_1 \leq \eta_2 \leq \cdots \leq \eta_N = \lambda_J$. From (9), it immediately follows that the error

$$\tilde{e}^n = \tilde{u} - \tilde{u}^n = \sum_{i=1}^{N} \tilde{e}_i^n \tilde{\psi}_i$$

satisfies

$$\tilde{e}_i^{n+1} = (1 - \eta_i / \underset{\approx}{\lambda}_J) \tilde{e}_i^n.$$

This means that the rate of reduction of the i'th error component is $\rho_i = (1 - \eta_i / \underset{\approx}{\lambda}_J)$. For η_i on the order of $\underset{\approx}{\lambda}_J$, this reduction is bounded away from one and corresponds to a good reduction. However, for η_i near $\eta_1 \approx ch^2$,

$\rho_i = 1 - ch^2$ which is near 1. This means that the components of the error corresponding to small eigenvalues will converge to zero rather slowly.

The two-level multigrid algorithm can now be motivated as follows. Since the form $(A_J\tilde{v}, \tilde{v})$ is equivalent to the sum of the squares of differences between neighboring mesh points, the components corresponding to the larger eigenvalues are highly oscillatory. In contrast, the components corresponding to the smaller eigenvalues are smoother and should be adequately approximated by coarser grid functions. This suggests combining a simple iteration of the form of (9) (to reduce the components corresponding to large eigenvalues) with a coarse grid solve (to reduce the components corresponding to smaller eigenvalues).

To describe this procedure, it is most natural to consider the computational problem more abstractly as one of finding functions in finite element subspaces. To this end, we define linear operators on the finite element spaces as follows: Let $A_k : M_k \to M_k$ be defined by

$$(A_k v, \phi) = D(v, \phi), \qquad \text{for all } \phi \in M_k. \tag{10}$$

Clearly, A_k is well defined since M_k is finite dimensional. Moreover, A_k is clearly symmetric and positive definite with respect to the inner product $(\cdot, \cdot)$. Let Q_k denote the $L^2(\Omega)$ projection onto M_k and let P_k denote the orthogonal projector with respect to the $D(\cdot, \cdot)$ inner product. These are defined by

$$(Q_k v, \phi) = (v, \phi), \qquad \text{for all } v \in L^2(\Omega),\ \phi \in M_k,$$

and

$$D(P_k v, \phi) = D(v, \phi), \qquad \text{for all } v \in H^1(\Omega),\ \phi \in M_k.$$

Now (4) can be rewritten as

$$A_J u = f_J \equiv Q_J f.$$

We denote by λ_J the largest eigenvalue of A_J. By the well known inverse properties for M,

$$(A_J v, v) = D(v, v) \leq Ch^{-2}(v, v), \qquad \text{for all } v \in M.$$

This means that the largest eigenvalue λ_J of A_J is bounded by Ch^{-2}. The Poincaré inequality implies that the smallest eigenvalue of A_J is bounded from below.

In this notation, the two-level multigrid algorithm is given as follows.

Algorithm 1.1 *Given $u^i \in M$ approximating the solution $u = A_J^{-1} f_J$ of (4), define $u^{i+1} \in M$ as follows:*

(1) *Set $u^{i+1/3} = u^i + \lambda_J^{-1}(f_J - A_J u^i)$.*
(2) *Define $u^{i+2/3} = u^{i+1/3} + q$, where*

$$A_{J-1} q = Q_{J-1}(f_J - A_J u^{i+1/3}).$$

(3) *Finally, set* $u^{i+1} \equiv u^{i+3/3} = u^{i+2/3} + \lambda_J^{-1}(f_J - A_J u^{i+2/3})$.

Steps 1 and 3 above are simple iteration of the form

$$u^{i+1} = u^i + \lambda_J^{-1}(f_J - A_J u^i). \tag{11}$$

In terms of the coefficient vectors $\tilde{u}^i$ and the dual vector $\underset{\sim}{f}_J$, it can be written as

$$\tilde{u}^{i+1} = \tilde{u}^i + \lambda_J^{-1} \underset{\approx}{G}_J^{-1}(\underset{\sim}{f}_J - \underset{\approx}{A}_J \tilde{u}^i)$$

where $\underset{\approx}{G}_J = [(\phi_i, \phi_j)]$ is the Gram matrix. This matrix is symmetric and positive definite and all its eigenvalues are on the order of Ch_J^2. Note that $\lambda_J \approx h_J^{-2}$ and $\lambda_J \approx 1$, hence $\lambda_J^{-1} \underset{\approx}{G}_J^{-1}$ is spectrally equivalent to $\underset{\approx}{\lambda}_J^{-1} I$. Consequently iteration (11), although slightly different from iteration (9), has a smoothing property similar to that of (9), i.e., the resulting error after application of (11) should be less oscillatory. Steps 1 and 3 above are often referred to as smoothing steps.

The middle step is called a coarse grid correction step. Note that for $\phi \in M_{J-1}$,

$$(Q_{J-1} A_J v, \phi) = (A_J v, \phi) = D(v, \phi) = D(P_{J-1} v, \phi) = (A_{J-1} P_{J-1} v, \phi),$$

i.e., $Q_{J-1} A_J = A_{J-1} P_{J-1}$. Set $e^i = u - u^i$ and $e^{i+j/3} = u - u^{i+j/3}$. We see that

$$q = A_{J-1}^{-1} Q_{J-1} A_J (u - u^{i+1/3}) = P_{J-1} e^{i+1/3}.$$

Thus, $e^{i+2/3} = (I - P_{J-1}) e^{i+1/3}$. This means that $e^{i+2/3}$ is the $D(\cdot,\cdot)$ orthogonal projection of $e^{i+1/3}$ into the subspace of M_J which is orthogonal to M_{J-1}. We now prove the following theorem.

Theorem 1.2 *Set* $e^i = u - u^i$, *where* u *is the solution of* (4) *and* u^i *is given by Algorithm 1.1. Then*

$$|||e^{i+1}||| \le \delta |||e^i|||.$$

Here $\delta < 1$ *independently of* h *and* $|||\cdot|||$ *denotes the norm defined by*

$$|||v||| = D(v, v)^{1/2}.$$

Proof. From Step (1) and Step (3) of Algorithm 1.1 and the above discussion,

$$e^{i+1/3} = (I - \lambda_J^{-1} A_J) e^i,$$
$$e^{i+2/3} = (I - P_{J-1}) e^{i+1/3},$$

and

$$e^{i+1} = e^{i+3/3} = (I - \lambda_J^{-1} A_J) e^{i+2/3}.$$

Thus,

$$e^{i+1} = (I - \lambda_J^{-1} A_J)(I - P_{J-1})(I - \lambda_J^{-1} A_J) e^i \equiv \mathcal{E} e^i.$$

Clearly,

$$|||e^{i+1}||| \leq |||\mathcal{E}|||\ |||e^i|||,$$

where $|||\mathcal{E}||| \equiv \sup_{v\in M_J} |||\mathcal{E}v|||/|||v|||$ is the operator norm of $\mathcal{E}$. Note that $\mathcal{E}$ is symmetric with respect to $D(\cdot,\cdot)$. Hence

$$\begin{aligned}
|||\mathcal{E}||| &= \sup_{v\in M_J} \frac{|D(\mathcal{E}v, v)|}{|||v|||^2} \\
&= \sup_{v\in M_J} \frac{|||(I-P_{J-1})(I-\lambda_J^{-1}A_J)v|||^2}{|||v|||^2} \\
&= |||(I-P_{J-1})(I-\lambda_J^{-1}A_J)|||^2 \\
&= |||(I-\lambda_J^{-1}A_J)(I-P_{J-1})|||^2.
\end{aligned}$$

We will estimate the last norm above. Let $w \in M$ and set $\hat{w} = (I-P_{J-1})w$. Then $D(\hat{w},\theta) = 0$ for all $\theta \in M_{J-1}$. Hence, for any $\theta \in M_{J-1}$,

$$D(\hat{w},\hat{w}) = D(\hat{w},\hat{w}-\theta) = (A_J\hat{w},\hat{w}-\theta) \leq (A_J\hat{w},A_J\hat{w})^{1/2}\|\hat{w}-\theta\|_0.$$

Using the approximation property (5) gives

$$D(\hat{w},\hat{w}) \leq Ch_{J-1}(A_J\hat{w},A_J\hat{w})^{1/2}D(\hat{w},\hat{w})^{1/2}.$$

Cancelling the common factor and using $\lambda_J \leq Ch_J^{-2} = 4Ch_{J-1}^{-2}$, we obtain,

$$D(\hat{w},\hat{w}) \leq Ch_{J-1}^2(A_J\hat{w},A_J\hat{w}) \leq \hat{C}\lambda_J^{-1}(A_J\hat{w},A_J\hat{w}).$$

Now $I-\lambda_J^{-1}A_J$ is symmetric with respect to $D(\cdot,\cdot)$ and $\sigma(I-\lambda_J^{-1}A_J) \subseteq [0,1)$. Hence

$$\begin{aligned}
|||(I-\lambda_J^{-1}A_J)\hat{w}|||^2 &\leq D((I-\lambda_J^{-1}A_J)\hat{w},\hat{w}) \\
&= D(\hat{w},\hat{w}) - \lambda_J^{-1}D(A_J\hat{w},\hat{w}) \\
&\leq (1-1/\hat{C})D(\hat{w},\hat{w}) \\
&\leq (1-1/\hat{C})D(w,w) = (1-1/\hat{C})|||w|||^2.
\end{aligned}$$

This shows that $|||\mathcal{E}||| = |||(I-\lambda_J^{-1}A_J)(I-P_{J-1})|||^2 \leq (1-1/\hat{C})$. Therefore

$$|||e^{i+1}||| \leq (1-1/\hat{C})\ |||e^i||| = \delta\ |||e^i|||,$$

where $\delta = 1-1/\hat{C}$ is independent of h. □

Remark 1.1 *If we omit Step 1 in Algorithm 1.1, then*

$$e^{i+1} = (I-\lambda_J^{-1}A_J)(I-P_{J-1})e^i$$

and hence

$$
\begin{aligned}
|||e^{i+1}||| &= |||(I - \lambda_J^{-1} A_J)(I - P_{J-1}) e^i||| \\
&\le |||(I - \lambda_J^{-1} A_J)(I - P_{J-1})|||\ |||e^i||| \\
&\le \delta^{1/2}\ |||e^i|||.
\end{aligned}
$$

We obtain the same bound if we omit, instead, Step (3). One obvious advantage of the symmetric algorithm is that it can be used as a preconditioner in the PCG algorithm.

What have we gained by using the two-level scheme? In this example,

$$
\dim M_{J-1} \approx \frac{1}{4} \dim M_J.
$$

Hence A_{J-1}^{-1} is "cheaper" to compute than A_J^{-1}. This is not a significant improvement for very large problems. This algorithm was described only to illustrate the multilevel technique. More efficient algorithms will be developed in later sections by recursive application of the above idea.

Remark 1.2 *The algorithm described above is not too convenient for direct application. First of all, the largest eigenvalue explicitly appears in Steps 1 and 3. It is not difficult to see that the largest eigenvalue in these algorithms can be replaced by any upper bound λ'_J for λ_J provided that $\lambda'_J \le C\lambda_J$. More importantly, Steps 1 and 3 require that Gram matrix systems be solved at each step of the iteration. Even though these matrices are well conditioned, this results in some unnecessary computational effort. Both of the above mentioned problems will be avoided by the introduction of more appropriate smoothing procedures. Smoothing procedures are studied in more detail in [3]. The more abstract multigrid algorithms of which we will now introduce use generic smoothing operators in place of $\lambda_J^{-1} I$.*

2 Multigrid I

In this section we develop the abstract multilevel theory. The first part generalizes the two-level method of the introduction to the multilevel case and an analysis of the V-cycle algorithm is given under somewhat restrictive conditions. The first convergence theory for the V-cycle is given in the case in which the multilevel spaces are nested and the corresponding forms are inherited from the form on the finest space. The case in which smoothing is done only on certain subspaces is not treated here but may be found in [3]. That theory covers the case of local mesh refinements discussed in [3] in the chapter on applications. A more general setting is introduced in next section of these notes. There, the requirements that the spaces be nested and that the forms be inherited is dropped. The W-cycle is defined as well as the variable V-cycle preconditioner.

2.1 An Abstract V-cycle Algorithm

We consider an abstract V-cycle algorithm in this section. We shall provide a theorem which can be used to estimate the rate of convergence of this multigrid algorithm. The analysis follows that of [7] and requires the so-called *full regularity and approximation* assumption. The fundamental ideas of this analysis are contained in the paper [2]. However, multigrid methods are applied to many problems which do not satisfy this hypothesis. Further results may be found in [3] which give some bounds for the rate of convergence in the case of less than full regularity. The first theorem provides a uniform convergence estimate while the second gives rise to a rate of convergence which deteriorates with the number of levels. In the next section a stronger result is proved using a product representation of the error.

For generality, the algorithms and theory presented in this section are given abstractly. However, the usefulness of any abstract theory depends on the possibility of verification of the hypotheses in various applications. We will indicate how these conditions are satisfied for many applications. Again, details are contained in [3].

2.2 The Multilevel Framework

In this subsection, we will set up the abstract multilevel framework. Let us consider a nested sequence of finite dimensional vector spaces

$$M_1 \subset M_2 \subset \ldots \subset M_J.$$

In addition, let $A(\cdot,\cdot)$ and $(\cdot,\cdot)$ be symmetric positive definite bilinear forms on M_J. The norm corresponding to $(\cdot,\cdot)$ will be denoted by $\|\cdot\|$. We shall develop multigrid algorithms for the solution of the following problem: Given $f \in M_J$, find $u \in M_J$ satisfying

$$A(u,\phi) = (f,\phi), \qquad \text{for all } \phi \in M_J. \tag{12}$$

As in the previous section, we will use auxiliary operators to define the multigrid algorithms. For $k = 1, \ldots, J$, let the operator $A_k : M_k \to M_k$ be defined by

$$(A_k v, \phi) = A(v,\phi), \qquad \text{for all } \phi \in M_k.$$

The operator A_k is clearly positive definite and symmetric in both the $A(\cdot,\cdot)$ and $(\cdot,\cdot)$ inner products. Also, we define the projectors $P_k : M_J \to M_k$ and $Q_k : M_J \to M_k$ by

$$A(P_k v, \phi) = A(v,\phi), \qquad \text{for all } \phi \in M_k,$$

and

$$(Q_k v, \phi) = (v,\phi), \qquad \text{for all } \phi \in M_k.$$

Note that (12) can be rewritten in the above notation as

$$A_J u = f. \tag{13}$$

We will use generic smoothing operators in our abstract multigrid algorithm. Assume that we are given linear operators, $R_k : M_k \to M_k$ for $k = 2, \ldots, J$. Denote by R_k^t the adjoint of R_k with respect to $(\cdot,\cdot)$. We shall state further conditions concerning these operators as needed in the discussion. The construction and analysis of effective smoothing operators may be found in [3].

2.3 The Abstract V-cycle Algorithm, I

We describe a V-cycle multigrid algorithm for computing the solution u of (13) by means of an iterative process. This involves recursively applying a generalization of Algorithm 1.1. Given an initial iterate $u^0 \in M_J$, we define a sequence approximating u by

$$u^{m+1} = \mathrm{Mg}_J(u^m, f). \tag{14}$$

Here $\mathrm{Mg}_J(\cdot,\cdot)$ is the map of $M_J \times M_J$ into M_J defined by the following algorithm.

Algorithm 2.1 *Set* $\mathrm{Mg}_1(v, g) = A_1^{-1} g$. *Let k be greater than one and let v and g be in M_k. Assuming that* $\mathrm{Mg}_{k-1}(\cdot,\cdot)$ *has been defined, we define* $\mathrm{Mg}_k(v, g)$ *as follows:*

(1) *Set* $v' = v + R_k^t(g - A_k v)$.
(2) *Set* $v'' = v' + q$ *where* $q = \mathrm{Mg}_{k-1}(0, Q_{k-1}(g - A_k v'))$.
(3) *Define* $\mathrm{Mg}_k(v, g) \equiv v''' = v'' + R_k(g - A_k v'')$.

The above algorithm is a recursive generalization of Algorithm 1.1. In fact, if we take $J = 2$ and $R_J = \lambda_J^{-1} I$ then Algorithm 2.1 coincides with Algorithm 1.1. The above algorithm replaces the solve on the $J-1$'st subspace (in Algorithm 1.1) with one recursive application of the multilevel procedure.

A straightforward induction argument shows that $\mathrm{Mg}_k(\cdot,\cdot)$ is a linear map of $M_k \times M_k$ into M_k. Moreover, it is obviously consistent. The linear operator $\mathcal{E}_J v \equiv \mathrm{Mg}_J(v, 0)$ is the error reduction operator for (14). That is

$$\mathcal{E}_J(u - u^m) \equiv u - u^{m+1} = \mathrm{Mg}_J(u, f) - \mathrm{Mg}_J(u^m, f) = \mathrm{Mg}_J(u - u^m, 0).$$

Steps 1 and 3 above are referred to as smoothing steps. Step 2 is called a coarse grid correction step. Step 3 is included so that the resulting linear multigrid operator

$$B_J g = \mathrm{Mg}_J(0, g)$$

is symmetric with respect to the inner product $(\cdot,\cdot)$, and hence can be used as a preconditioner for A_J. It is straightforward to show that with $B_k g = \mathrm{Mg}_k(0, g)$,

$$\mathrm{Mg}_k(v, g) = v + B_k(g - A_k v).$$

To use B_k as a preconditioner, it is often convenient to define $B_k g \equiv \mathrm{Mg}_k(0, g)$ directly and the following algorithm gives a recursive definition of B_k.

Algorithm 2.2 *Let $B_1 = A_1^{-1}$. Assuming $B_{k-1} : M_{k-1} \to M_{k-1}$ has been defined, we define $B_k : M_k \to M_k$ as follows. Let $g \in M_k$.*

(1) *Set $v' = R_k^t g$.*
(2) *Set $v'' = v' + q$ where $q = B_{k-1} Q_{k-1}(g - A_k v')$.*
(3) *Set $B_k g \equiv v''' = v'' + R_k(g - A_k v'')$.*

2.4 The Two-level Error Recurrence

We next derive a two-level error recurrence for the multigrid process defined by Algorithm 2.1, i.e., we compute an expression for the error operator $\mathcal{E}_k v = \mathrm{Mg}_k(v, 0)$. Let $u = A_k^{-1} g$. By consistency, $\mathcal{E}_k(u-v) \equiv u - \mathrm{Mg}_k(v, g) = \mathrm{Mg}_k(u-v, 0)$.

We want to express $u - \mathrm{Mg}_k(v, g)$ in terms of $u - v$. We start by considering the effect of the smoothing steps.

Now v''' of Algorithm 2.1 satisfies

$$u - \mathrm{Mg}_k(v, g) = u - v''' = (I - R_k A_k)(u - v'') \equiv K_k(u - v'') \tag{15}$$

and v' satisfies

$$u - v' = (I - R_k^t A_k)(u - v) \equiv K_k^*(u - v) \tag{16}$$

Note that $K_k^* = (I - R_k^t A_k)$ is the adjoint of K_k with respect to the $A(\cdot, \cdot)$ inner product.

We next consider the effect of the coarse grid correction step. Note that for $\phi \in M_i$ with $i < k$,

$$(Q_i A_k v, \phi) = (A_k v, \phi) = A(v, \phi) = A(P_i v, \phi) = (A_i P_i v, \phi).$$

This means that $Q_i A_k = A_i P_i$. Thus, q of Step 2 is given by

$$\begin{aligned} q &= \mathrm{Mg}_{k-1}(0, Q_{k-1}(g - A_k v')) \\ &= \mathrm{Mg}_{k-1}(0, Q_{k-1} A_k(u - v')) \\ &= \mathrm{Mg}_{k-1}(0, A_{k-1} P_{k-1}(u - v')) \\ &= P_{k-1}(u - v') - \mathcal{E}_{k-1} P_{k-1}(u - v'). \end{aligned}$$

We used consistency for the last equality. Consequently,

$$u - v'' = [(I - P_{k-1}) + \mathcal{E}_{k-1} P_{k-1}](u - v'). \tag{17}$$

Combining (15), (16) and (17) gives the two-level error recurrence

$$\mathcal{E}_k = K_k[(I - P_{k-1}) + \mathcal{E}_{k-1} P_{k-1}] K_k^*. \tag{18}$$

A simple argument using mathematical induction shows that $\mathcal{E}_k$ is symmetric, positive semidefinite with respect to the $A(\cdot,\cdot)$ inner product, i.e.,

$$0 \le A(\mathcal{E}_k v, v), \qquad \text{for all } v \in M_k.$$

In particular,

$$0 \le A(\mathcal{E}_J v, v), \qquad \text{for all } v \in M_J. \tag{19}$$

2.5 The Braess-Hackbusch Theorem

We prove a result given by [2] in this subsection. This result was the first which showed that multigrid algorithms could be expected to converge with very weak conditions on the smoother.

For this result, we require two conditions. The first is on the smoother. Let λ_k be the largest eigenvalue of A_k. Let ω be in $(0,2)$. For $v, g \in M_k$, Richardson's iteration for the solution of $A_k v = g$ is defined by

$$v^{m+1} = v^m + \omega\lambda_k^{-1}(g - A_k v^m).$$

The corresponding error reduction operator is

$$K_{k,\omega} = I - (\omega\lambda_k^{-1})A_k.$$

Note that $K_k = (I - R_k A_k)$ and

$$A(K_k v, K_k v) = A((I - \overline{R}_k A_k)v, v),$$

where

$$\overline{R}_k = (I - K_k^* K_k)A_k^{-1} = R_k^t + R_k - R_k^t A_k R_k$$

is the linear smoothing operator corresponding to the symmetrized iterative process with reducer $K_k^* K_k$. Our condition concerning the smoothers R_k, $k = 2, \dots, J$ is related to Richardson's iteration with an appropriate parameter ω.

Condition (SM.1) There exists an $\omega \in (0,1)$, not depending on J, such that

$$0 \le A(K_k v, K_k v) \le A(K_{k,\omega} v, v), \qquad \text{for all } v \in M_k,\ k = 2, 3, \dots, J,$$

or equivalently

$$\frac{\omega}{\lambda_k}(v, v) \le (\overline{R}_k v, v) \le (A_k^{-1} v, v), \qquad \text{for all } v \in M_k,\ k = 2, 3, \dots, J.$$

Remark 2.1 *Note that both $K_k^* K_k$ and $K_{k,\omega}$ are symmetric with respect to $A(\cdot,\cdot)$ and positive semi-definite on M_k. Condition (SM.1) implies that $K_k^* K_k$ is less than or equal to $K_{k,\omega}$. That is, the smoother $\overline{R}_k$ results in a reduction which is at least as good as the result obtained using the Richardson smoother $(\omega/\lambda_k)I$.*

The second condition which we will require is the so-called *full regularity and approximation* condition.

Condition (A.1) There is a constant C_{P} not depending on J such that

$$\|(I - P_{k-1})v\|^2 \le C_{\mathrm{P}} \lambda_k^{-1} A(v, v), \qquad \text{for all } v \in M_k,\ k = 2, \ldots, J.$$

In applications, the verification of this condition requires that the underlying elliptic problem satisfy full elliptic regularity. In the model problem of Section 1, P_{k-1} is the elliptic projector. The above condition is obtained by application of finite element duality techniques.

We can now state the Braess-Hackbusch Theorem.

Theorem 2.1 *Let* $\mathrm{Mg}_J(\cdot, \cdot)$ *be defined by Algorithm 2.1. Assume that Condition* (A.1) *holds and that the smoothers* R_k *satisfy Condition* (SM.1). *Then the error reduction operator* $\mathcal{E}_J v = \mathrm{Mg}_J(v, 0)$ *of Algorithm 2.1 satisfies*

$$0 \le A(\mathcal{E}_J v, v) \le \frac{C_{\mathrm{P}}}{C_{\mathrm{P}} + \omega} A(v, v), \qquad \textit{for all } v \in M_J.$$

This means that the sequence u^m *defined by* (14) *satisfies*

$$|||u - u^m||| \le \left(\frac{C_{\mathrm{P}}}{C_{\mathrm{P}} + \omega} \right)^m |||u - u^0|||.$$

Before proving this theorem, we state and prove the following lemma.

Lemma 2.1 *Assume that Condition* (SM.1) *holds. Then for any* $v \in M_k$,

$$\frac{1}{\lambda_k} \|A_k K_k^* v\|^2 \le \frac{1}{\omega} [A(v, v) - A(K_k^* v, K_k^* v)].$$

Proof. Let v be in M_k. By Condition (SM.1) with v replaced by $A_k K_k^* v$,

$$\begin{aligned} \frac{1}{\lambda_k} \|A_k K_k^* v\|^2 &\le \frac{1}{\omega} (\overline{R}_k A_k K_k^* v, A_k K_k^* v) = \frac{1}{\omega} A((I - K_k^* K_k) K_k^* v, K_k^* v) \\ &= \frac{1}{\omega} A((I - \bar{K}_k) \bar{K}_k v, v), \end{aligned}$$

where $\bar{K}_k = K_k K_k^*$. The operator $\bar{K}_k$ is symmetric and non-negative with respect to the $A(\cdot, \cdot)$ inner product.

It follows from the symmetry of $\bar{K}_k$ that

$$A((I - \bar{K}_k) \bar{K}_k v, v) \le A((I - \bar{K}_k) v, v) = [A(v, v) - A(K_k^* v, K_k^* v)].$$

Combining the above inequalities completes the proof of the lemma. □

Proof of Theorem 2.1. By (19), we need only prove the upper inequality. The proof is by induction. We will show that for $i = 1, \ldots, J$,

$$A(\mathcal{E}_i v, v) \le \delta A(v, v), \qquad \text{for all } v \in M_i \tag{20}$$

holds for $\delta = C_P/(C_P + \omega)$. Since $\mathcal{E}_1 = 0$ is the trivial operator on M_1, (20) obviously holds for $i = 1$. Let k be greater than one and assume that (20) holds for $i = k-1$. Let v be in M_k. Using the two-level recurrence formula (18) gives

$$A(\mathcal{E}_k v, v) = A((I - P_{k-1})K_k^* v, K_k^* v) + A(\mathcal{E}_{k-1} P_{k-1} K_k^* v, P_{k-1} K_k^* v).$$

Applying the induction hypothesis we have

$$\begin{aligned} A(\mathcal{E}_k v, v) &\le A((I - P_{k-1})K_k^* v, K_k^* v) + \delta A(P_{k-1} K_k^* v, P_{k-1} K_k^* v) \\ &= (1-\delta) A((I - P_{k-1})K_k^* v, K_k^* v) + \delta A(K_k^* v, K_k^* v). \end{aligned} \tag{21}$$

By the Cauchy-Schwarz inequality and Condition (A.1),

$$\begin{aligned} A((I - P_{k-1})K_k^* v, K_k^* v) &\le \|(I - P_{k-1})K_k^* v\| \|A_k K_k^* v\| \\ &\le \left(C_P \lambda_k^{-1} A((I - P_{k-1})K_k^* v, K_k^* v)\right)^{1/2} \|A_k K_k^* v\|. \end{aligned}$$

Obvious manipulations give

$$A((I - P_{k-1})K_k^* v, K_k^* v) \le \frac{C_P}{\lambda_k} \|A_k K_k^* v\|^2. \tag{22}$$

Thus, (21), (22) and Lemma 2.1 imply that

$$\begin{aligned} A(\mathcal{E}_k v, v) &\le \frac{(1-\delta) C_P}{\omega} [A(v,v) - A(K_k^* v, K_k^* v)] + \delta A(K_k^* v, K_k^* v) \\ &= \delta A(v,v). \end{aligned}$$

This completes the proof of the theorem. □

Remark 2.2 *The original result given by [2] was stated in terms of the particular smoother $R_k = \lambda_k^{-1} I$. They also allowed a fixed number m of applications of this smoother on each level. Theorems with variable numbers of smoothings will appear later in these notes.*

Results in this section are based on the two-level error recurrence for the multigrid algorithm. We refer to the survey paper [4] for more detailed discussion on the development of the multigrid methods.

3 Multigrid II: V-cycle with Less Than Full Elliptic Regularity

3.1 Introduction and Preliminaries

In this section, we provide an analysis of Algorithm 2.1 with an assumption which is weaker than Condition (A.1). To state this condition, we require scales of discrete norms. The operator A_k is symmetric and positive definite on M_k. Consequently, its fractional powers are well defined. For real s, we consider the scale of discrete norms on M_k defined by $|||\cdot|||_{s,k}$

$$|||v|||_{s,k} = (A_k^s v, v)^{1/2}, \qquad \text{for all } v \in M_k.$$

Note that

$$|||v|||_{0,k} = \|v\| \quad \text{and} \quad |||v|||_{1,k} = |||v|||.$$

By expanding in the eigenfunctions of A_k and using the Hölder inequality, it follows easily that

$$|||v|||_{1+\alpha,k} \le |||v|||_{2,k}^{\alpha}\, |||v|||_{1,k}^{1-\alpha}, \qquad 0 \le \alpha \le 1.$$

Our theorem will be based on the following generalization of Condition (A.1).

Condition (A.2) There exists a number $\alpha \in (0,1]$ and a constant C_{P} not depending on J such that

$$|||(I - P_{k-1})v|||_{1-\alpha,k}^2 \le C_{\mathrm{P}}^{\alpha}\, \lambda_k^{-\alpha}\, A(v,v), \qquad \text{for all } v \in M_k,\ k = 2, \ldots, J.$$

In the model problem of Section 1, P_{k-1} is the elliptic projector. It can be shown that for this example and $s \in [0,1]$, the norm $|||\cdot|||_{s,k}$ is equivalent on M_k (with constants independent of J) to the Sobolev norm $\|\cdot\|_s$. The above condition can be then obtained by application of finite element duality techniques.

We will also need a geometrical growth property of λ_k which can be verified easily for elliptic finite element problems on a sequence of nested meshes. We postulate this property in the following condition.

Condition (A.3) There exist positive constants $\gamma_1 \le \gamma_2 < 1$ such that

$$\gamma_1 \lambda_{k+1} \le \lambda_k \le \gamma_2 \lambda_{k+1} \quad \text{for} \quad k = 1, 2, \ldots, J-1.$$

When using this condition, it is convenient to introduce the symmetric matrix $\Lambda^{(\alpha)}$, with $0 < \alpha \le 1$, whose lower triangular entries are given by

$$\Lambda_{ki}^{(\alpha)} = (\lambda_k/\lambda_i)^{\alpha/2}, \quad \text{for } k \le i. \tag{23}$$

Denote by $\Lambda_L^{(\alpha)}$ and $\Lambda_U^{(\alpha)}$ the lower and upper triangular parts of $\Lambda^{(\alpha)}$. Condition (A.3) then implies that $\Lambda_{ki}^{(\alpha)} = (\lambda_k/\lambda_i)^{\alpha/2} \le \gamma_2^{(i-k)\alpha/2}$. Therefore,

$$\|\Lambda_L^{(\alpha)}\|_{\ell_2} = \|\Lambda_U^{(\alpha)}\|_{\ell_2} \le \left(\frac{\gamma_2^{\alpha/2}}{1-\gamma_2^{\alpha/2}}\right) \quad \text{and} \quad \|\Lambda^{(\alpha)}\|_{\ell_2} \le \left(\frac{1+\gamma_2^{\alpha/2}}{1-\gamma_2^{\alpha/2}}\right).$$

The last condition needed is a condition on the uniform boundedness of the operators Q_k. This is the following.

Condition (A.4) Let $\alpha \in (0,1]$ be given. Then there exists a constant C'_{Q} not depending on J such that

$$|||Q_k v|||^2_{1-\alpha,k} \le C'_{\mathrm{Q}} |||v|||^2_{1-\alpha,i}, \qquad \text{for all } v \in M_i,\ 1 \le k \le i \le J.$$

Before proceeding we will prove a basic lemma.

Lemma 3.1 *Assume that Conditions* (A.2) *and* (A.4) *hold for the same* $\alpha \in (0,1]$ *and that Condition* (A.3) *holds. Then*

$$\sum_{k=1}^{J} |||(P_k - Q_k)v|||^2 \le C'_b A(v,v), \qquad \text{for all } v \in M_J, \tag{24}$$

$$A(v,v) \le C'_c \left(|||Q_1 v|||^2 + \sum_{k=2}^{J} |||(Q_k - Q_{k-1})v|||^2 \right), \qquad \text{for all } v \in M_J, \tag{25}$$

and

$$|||Q_1 v|||^2 + \sum_{k=2}^{J} \lambda_k \|(Q_k - Q_{k-1})v\|^2 \le C'_a A(v,v), \qquad \text{for all } v \in M_J. \tag{26}$$

Here $C'_b = (C'_{\mathrm{Q}} C^{\alpha}_{\mathrm{P}}) \left(\frac{\gamma_2^{\alpha/2}}{1-\gamma_2^{\alpha/2}} \right)^2$, $C'_c = (\sqrt{C'_b} + \sqrt{C'_b + 1})^2$ *and* $C'_a = 4C'_{\mathrm{Q}} C_{\mathrm{P}} \left(\frac{1}{1-\gamma_2^{\alpha/2}} \right)^2$.

Proof. Let v be in M_J. By the definition of λ_k,

$$\begin{aligned} \sum_{k=1}^{J} |||(P_k - Q_k)v|||^2 &\le \sum_{k=1}^{J} \left(\lambda_k^{\alpha/2} |||(P_k - Q_k)v|||_{1-\alpha,k} \right)^2 \\ &= \sum_{k=1}^{J} \left(\lambda_k^{\alpha/2} |||Q_k(P_k - I)v)|||_{1-\alpha,k} \right)^2. \end{aligned}$$

Note that $(I - P_k) = \sum_{i=k+1}^{J} (P_i - P_{i-1})$. By the triangle inequality and Condition (A.4),

$$\sum_{k=1}^{J} |||(P_k - Q_k)v|||^2 \le C'_{\mathrm{Q}} \sum_{k=1}^{J} \left(\sum_{i=k+1}^{J} \lambda_k^{\alpha/2} |||(P_i - P_{i-1})v|||_{1-\alpha,i} \right)^2.$$

Note that $(P_i - P_{i-1})$ is a projector and hence Condition (A.2) implies that

$$\sum_{k=1}^{J} |||(P_k - Q_k)v|||^2 \le C'_{\mathrm{Q}} C_{\mathrm{P}}^{\alpha} \sum_{k=1}^{J} \left(\sum_{i=k+1}^{J} \left(\frac{\lambda_k}{\lambda_i}\right)^{\alpha/2} |||(P_i - P_{i-1})v||| \right)^2$$
$$\le C'_{\mathrm{Q}} C_{\mathrm{P}}^{\alpha} \|\Lambda_U^{(\alpha)}\|_{\ell_2}^2 \sum_{k=2}^{J} |||(P_k - P_{k-1})v|||^2.$$

Condition (A.3) implies that $\|\Lambda_U^{(\alpha)}\|_{\ell_2}^2 \le \left(\frac{\gamma_2^{\alpha/2}}{1-\gamma_2^{\alpha/2}}\right)^2$. Inequality (24) now follows from the orthogonality of $(P_i - P_{i-1})$.

Now let $v_k = (Q_k - Q_{k-1})v$. Then $\sum_{i=k+1}^{J} v_i = (I - Q_k)v$. We have

$$\begin{aligned}
A(v,v) &= \sum_{k,i=1}^{J} A(v_k, v_i) = \sum_{k=1}^{J} A(v_k, v_k) + 2\sum_{k=1}^{J}\sum_{i=k+1}^{J} A(v_k, v_i) \\
&= \sum_{k=1}^{J} A(v_k, v_k) + 2\sum_{k=1}^{J} A(v_k, (I - Q_k)v) \\
&= \sum_{k=1}^{J} A(v_k, v_k) + 2\sum_{k=1}^{J} A(v_k, (P_k - Q_k)v) \\
&\le \sum_{k=1}^{J} |||v_k|||^2 + 2\left(\sum_{k=1}^{J} |||v_k|||^2\right)^{1/2} \left(\sum_{k=1}^{J} |||(P_k - Q_k)v|||^2\right)^{1/2} \\
&\le \sum_{k=1}^{J} |||v_k|||^2 + 2\left(\sum_{k=1}^{J} |||v_k|||^2\right)^{1/2} [C'_b A(v,v)]^{1/2}.
\end{aligned}$$

Elementary algebra shows that

$$A(v,v) \le C'_c \sum_{k=1}^{J} |||v_k|||^2 = C'_c \left(|||Q_1 v|||^2 + \sum_{k=2}^{J} |||(Q_k - Q_{k-1})v|||^2 \right)$$

for $C'_c = (\sqrt{C'_b} + \sqrt{C'_b + 1})^2$. This proves (25).

For a proof of the (26) see [3]. □

We now return to the multigrid method defined by Algorithm 2.1. In this section we continue in the framework of the previous two sections and establish four additional theorems which provide bounds for the convergence rate of Algorithm 2.1. The proofs of these results depend on expressing the error propagator for the multigrid algorithm in terms of a product of operators defined on the finest space. The theorem presented here gives a uniform convergence estimate and can be used in many cases without full elliptic regularity. Several other results based on this error representation may be found in [3].

3.2 The Multiplicative Error Representation

The theory of Subsection 2.1 was based on a two-level error representation. To get strong results in the case of less than full elliptic regularity, we shall need to express the error in a different way. The fine grid error operator $\mathcal{E}_J$ is expressed as a product of factors associated with the smoothings on individual levels.

By (18), for $k > 1$ and any $v \in M_J$,

$$(I - P_k)v + \mathcal{E}_k P_k v = (I - P_k)v + K_k[(I - P_{k-1}) + \mathcal{E}_{k-1}P_{k-1}]K_k^* P_k v.$$

Let $T_k = R_k A_k P_k$ for $k > 1$ and set $T_1 = P_1$. The adjoint of T_k (for $k > 1$) with respect to the inner product $A(\cdot,\cdot)$ is given by $T_k^* = R_k^t A_k P_k$. Note that $I - T_k$ and $I - T_k^*$ respectively extend the operators K_k and K_k^* to operators defined on all of M_J. Clearly $(I - P_k)v = 0$ for all $v \in M_k$. In addition, P_k commutes with T_k. Thus,

$$\begin{aligned}(I - P_k)v + \mathcal{E}_k P_k v &= (I - T_k)[(I - P_{k-1}P_k) + \mathcal{E}_{k-1}P_{k-1}P_k](I - T_k^*)v \\ &= (I - T_k)[(I - P_{k-1}) + \mathcal{E}_{k-1}P_{k-1}](I - T_k^*)v.\end{aligned}$$

The second equality follows from the identity $P_{k-1}P_k = P_{k-1}$. Repeatedly applying the above identity gives

$$\mathcal{E}_J v = (I - T_J)(I - T_{J-1}) \cdots (I - T_1)(I - T_1^*)(I - T_2^*) \cdots (I - T_J^*). \quad (27)$$

Equation (27) displays the J'th level error operator as a product of factors. These factors show the precise effect of the individual grid smoothings on the fine grid error propagator.

Our goal is to show that

$$0 \le A(\mathcal{E}_J v, v) \le (1 - 1/C_M)A(v,v). \quad (28)$$

This means that the sequence u^m defined by (14) satisfies

$$|||u - u^m||| \le (1 - 1/C_M)^m |||u - u^0|||.$$

Let $E_0 = I$ and set $E_k = (I - T_k)E_{k-1}$, for $k = 1, 2, \ldots, J$. By (27), $\mathcal{E}_J = E_J E_J^*$ where E_J^* is the adjoint of E_J with respect to the $A(\cdot,\cdot)$ inner product. We can rewrite (28) as

$$A(E_J^* v, E_J^* v) \le (1 - 1/C_M)A(v,v), \qquad \text{for all } v \in M_J,$$

which is equivalent to

$$A(E_J v, E_J v) \le (1 - 1/C_M)A(v,v), \qquad \text{for all } v \in M_J.$$

Consequently, (28) is equivalent to

$$A(v,v) \le C_M[A(v,v) - A(E_J v, E_J v)], \qquad \text{for all } v \in M_J. \quad (29)$$

We will establish a typical uniform convergence estimate for the multigrid algorithm by proving (29) using the above conditions. For some other results in this direction see [3].

3.3 Some Technical Lemmas

We will prove in this subsection an abstract estimate which is essential to all of our estimates. For this, we will need to use the following lemmas. The first two are simply identities. Recall that $\overline{R}_k = R_k + R_k^t - R_k^t A_k R_k$.

Lemma 3.2 *The following identity holds.*

$$A(v,v) - A(E_J v, E_J v) = \sum_{k=1}^{J} (\overline{R}_k A_k P_k E_{k-1} v, A_k P_k E_{k-1} v). \tag{30}$$

Proof. For any $w \in M_J$, a simple calculation shows that

$$A(w,w) - A((I-T_k)w, (I-T_k)w) = A((2I - T_k)w, T_k w).$$

Let v be in M_J. Taking $w = E_{k-1}v$ in the above identity and summing gives

$$\begin{aligned} A(v,v) - A(E_J v, E_J v) &= \sum_{k=1}^{J} A((2I - T_k)E_{k-1}v, T_k E_{k-1} v) \\ &= \sum_{k=1}^{J} (\overline{R}_k A_k P_k E_{k-1} v, A_k P_k E_{k-1} v). \end{aligned}$$

This is (30). □

Lemma 3.3 *Let $\pi_k : M \to M_k$ be a sequence of linear operators with $\pi_J = I$. Set $\pi_0 = 0$ and $E_0 = I$. Then the following identity holds.*

$$\begin{aligned} A(v,v) &= \sum_{k=1}^{J} A(E_{k-1}v, (\pi_k - \pi_{k-1})v) + \sum_{k=1}^{J-1} A(T_k E_{k-1} v, (P_k - \pi_k)v) \\ &= \left(A(P_1 v, v) + \sum_{k=2}^{J} A(E_{k-1}v, (\pi_k - \pi_{k-1})v) \right) \\ &\quad + \left(\sum_{k=2}^{J-1} A(T_k E_{k-1} v, (P_k - \pi_k)v) \right). \end{aligned} \tag{31}$$

Proof. Expanding v in terms of π_k gives

$$\begin{aligned} A(v,v) &= \sum_{k=1}^{J} A(v, (\pi_k - \pi_{k-1})v) \\ &= \sum_{k=1}^{J} A(E_{k-1}v, (\pi_k - \pi_{k-1})v) + \sum_{k=1}^{J} A((I - E_{k-1})v, (\pi_k - \pi_{k-1})v). \end{aligned}$$

Let $F_k = I - E_k = \sum_{i=1}^{k} T_i E_{i-1}$. Then since $F_0 = 0$ and $\pi_0 = 0$,

$$
\begin{aligned}
\sum_{k=1}^{J} & A(F_{k-1}v, (\pi_k - \pi_{k-1})v) \\
&= \sum_{k=1}^{J} A(F_{k-1}v, \pi_k v) - \sum_{k=1}^{J-1} A(F_k v, \pi_k v) \\
&= \sum_{k=1}^{J-1} A((F_{k-1} - F_k)v, \pi_k v) + A(F_{J-1}v, v).
\end{aligned}
$$

Now $F_{k-1} - F_k = -T_k E_{k-1}$. Rearranging the terms gives

$$
\begin{aligned}
\sum_{k=1}^{J} & A(F_{k-1}v, (\pi_k - \pi_{k-1})v) \\
&= - \sum_{k=1}^{J-1} A(T_k E_{k-1}v, \pi_k v) + \sum_{k=1}^{J-1} A(T_k E_{k-1}v, v) \\
&= \sum_{k=1}^{J-1} A(T_k E_{k-1}v, (P_k - \pi_k)v).
\end{aligned}
$$

This is the first equality in (31). Note that $R_1 = A_1^{-1}$ and $T_1 = P_1$ and thus the second equality in (31) follows from the first. □

Set $\|w\|_{\overline{R}_k^{-1}} = (\overline{R}_k^{-1} w, w)^{1/2}$ and $\|w\|_{\overline{R}_k} = (\overline{R}_k w, w)^{1/2}$ for all $w \in M_k$.

Using (30) and (31) we can prove the following basic estimate.

Lemma 3.4 *Let $\pi_k : M \to M_k$ be a sequence of linear operators with $\pi_J = I$. Let $\pi_0 = 0$. Then we have the following basic estimate.*

$$
A(v, v) \le \left[A(v, v) - A(E_J v, E_J v)\right]^{1/2} \left[\sigma_1(v) + \sigma_2(v)\right], \tag{32}
$$

where

$$
\sigma_1(v) = \left(A(P_1 v, v) + \sum_{k=2}^{J} \|(\pi_k - \pi_{k-1})v\|^2_{\overline{R}_k^{-1}} \right)^{1/2}
$$

and

$$
\sigma_2(v) = \left(\sum_{k=2}^{J-1} \|R_k^t A_k (P_k - \pi_k)v\|^2_{\overline{R}_k^{-1}} \right)^{1/2}.
$$

Proof. We will bound the right hand side of (31). By the Cauchy-Schwarz inequality, the first sum of (31) can be bounded as

$$
\begin{aligned}
A(P_1 v, v) &+ \sum_{k=2}^{J} A(E_{k-1}v, (\pi_k - \pi_{k-1})v) \\
&= A(P_1 v, v) + \sum_{k=2}^{J} (A_k P_k E_{k-1}v, (\pi_k - \pi_{k-1})v)
\end{aligned}
$$

$$\leq \left(\sum_{k=1}^{J} \|A_k P_k E_{k-1} v\|^2_{\overline{R}_k} \right)^{1/2} \left(A(P_1 v, v) + \sum_{k=2}^{J} \|(\pi_k - \pi_{k-1}) v\|^2_{\overline{R}_k^{-1}} \right)^{1/2}.$$

Similarly the second sum of (31) can be bounded as

$$\sum_{k=2}^{J-1} A(T_k E_{k-1} v, (P_k - \pi_k) v)$$

$$= \sum_{k=2}^{J-1} (R_k A_k P_k E_{k-1} v, A_k (P_k - \pi_k) v)$$

$$\leq \left(\sum_{k=2}^{J-1} \|A_k P_k E_{k-1} v\|^2_{\overline{R}_k} \right)^{1/2} \left(\sum_{k=2}^{J-1} \|R_k^t A_k (P_k - \pi_k) v\|^2_{\overline{R}_k^{-1}} \right)^{1/2}.$$

Combining these two estimates and using (31) we obtain

$$A(v,v) \leq \left(\sum_{k=1}^{J} (\overline{R}_k A_k P_k E_{k-1} v, A_k P_k E_{k-1} v) \right)^{1/2} [\sigma_1(v) + \sigma_2(v)].$$

Using (30) in the above inequality proves the lemma. □

3.4 Uniform Estimates

In this subsection, we will show that the uniform convergence estimate of Theorem 2.1 often extends to the case of less than full regularity and approximation. The results are obtained by bounding the second factor in Lemma 3.4 from above by $A(v, v)$.

We shall require some conditions on the smoother which we state here. Recall that $\overline{R}_k = R_k + R_k^t - R_k^t A_k R_k$, $K_k = (I - R_k A_k)$ and $K_{k,\omega} = I - (\omega \lambda_k^{-1}) A_k$.

Condition (SM.1) There exists a constant $\omega \in (0, 1)$ not depending on J such that

$$0 \leq A(K_k v, K_k v) \leq A(K_{k,\omega} v, v), \qquad \text{for all } v \in M_k,\ k = 2, 3, \ldots, J,$$

or equivalently

$$\frac{\omega}{\lambda_k}(v, v) \leq (\overline{R}_k v, v) \leq (A_k^{-1} v, v), \qquad \text{for all } v \in M_k,\ k = 2, 3, \ldots, J.$$

In addition to the smoother Condition (SM.1), we will require that the smoother R_k be properly scaled. More precisely the condition is the following:

Condition (SM.2) There exists a constant $\theta \in (0, 2)$ not depending on J such that

$$A(R_k v, R_k v) \le \theta (R_k v, v), \qquad \text{for all } v \in M_k,\ k = 2, 3, \ldots, J.$$

This inequality is the same as any one of the following three inequalities:

$$(\overline{R}_k v, v) \ge (2-\theta)(R_k v, v) \ge \left(\frac{2-\theta}{\theta}\right)((R_k^t A_k R_k)v, v), \quad \text{for all } v \in M_k.$$

A simple change of variable shows that Condition (SM.2) is equivalent to the following condition for $T_k = R_k A_k P_k$.

$$A(T_k v, T_k v) \le \theta A(T_k v, v), \qquad \text{for all } v \in M_J,\ k = 2, 3, \ldots, J.$$

Note that if the smoother R_k satisfies Condition (SM.1), then it also satisfies the inequality in Condition (SM.2) with $\theta < 2$, but possibly depending on k. Therefore smoothers that satisfy (SM.1) will satisfy Condition (SM.2) after a proper scaling.

Our theorem uses Condition (A.2), a regularity and approximation property of P_k, Condition (A.3), a geometrical growth condition for λ_k, and Condition (A.4), a stability condition on Q_k.

Theorem 3.1 *Assume that Conditions* (A.2) *and* (A.4) *hold with the same* α *and that Condition* (A.3) *holds. Assume in addition that the smoothers,* R_k*, satisfy Conditions* (SM.1) *and* (SM.2)*. Then,*

$$0 \le A(\mathcal{E}_J v, v) \le (1 - 1/C_M) A(v, v), \qquad \textit{for all } v \in M_J.$$

Here $C_M = \left[\left(\frac{\theta C_b'}{2-\theta}\right)^{1/2} + \left(1 + \frac{C_a'}{\omega}\right)^{1/2}\right]^2$ *with*

$$C_a' = 4 C_Q' C_P \left(\frac{1}{1-\gamma_2^{\alpha/2}}\right)^2 \quad \textit{and} \quad C_b' = (C_Q' C_P^{\alpha}) \left(\frac{\gamma_2^{\alpha/2}}{1-\gamma_2^{\alpha/2}}\right)^2..$$

Proof. The theorem is proved by bounding the second factor in the right hand side of (32) from above by $A(v, v)$. By Lemma 3.1,

$$\sum_{k=1}^{J} |||(P_k - Q_k)v|||^2 \le C_b' A(v, v), \qquad \text{for all } v \in M_J$$

and

$$|||Q_1 v|||^2 + \sum_{k=2}^{J} \lambda_k \|(Q_k - Q_{k-1})v\|^2 \le C_a' A(v, v), \qquad \text{for all } v \in M_J.$$

By Condition (SM.1), $(\overline{R}_k^{-1} w, w) \le \omega^{-1} \lambda_k \|w\|^2$, for all $w \in M_k$, $k \ge 2$. Hence,

$$A(P_1 v, v) + \sum_{k=2}^{J} \|(Q_k - Q_{k-1})v\|^2_{\overline{R}_k^{-1}} \le \left(1 + \frac{C_a'}{\omega}\right) A(v, v).$$

Condition (SM.2) implies that $(R_k \overline{R}_k^{-1} R_k^t w, w) \le \frac{\theta}{2-\theta}(A_k^{-1} w, w)$ for $w \in M_k$. Hence,

$$\sum_{k=2}^{J} \|R_k^t A_k (P_k - Q_k) v\|_{\overline{R}_k^{-1}}^2 \le \frac{\theta}{2-\theta} \sum_{k=2}^{J} |||(P_k - Q_k) v|||^2$$
$$\le \frac{\theta C_b'}{2-\theta} A(v,v).$$

Combining these two estimates and applying Lemma 3.4, with $\pi_k = Q_k$, shows that

$$A(v,v) \le C_M \, [A(v,v) - A(E_J v, E_J v)].$$

The theorem follows. □

4 Non-nested Multigrid

4.1 Non-nested Spaces and Varying Forms

There is not a lot in the literature on this subject. Since one of the lectures was devoted to this more general point of view, and an extensive treatment was given in [3], the material presented here is largely taken from [3].

We therefore depart from the framework of the previous sections where we considered only nested sequences of finite dimensional spaces. More generally, assume that we are given a sequence of finite dimensional spaces $M_1, M_2, \ldots, M_J = M$. Each space M_k is equipped with an inner product $(\cdot,\cdot)_k$ and denote by $\|\cdot\|_k$ the induced norm. In addition, we assume that we are given symmetric and positive definite forms $A_k(\cdot,\cdot)$ defined on $M_k \times M_k$ and set $A(\cdot,\cdot) = A_J(\cdot,\cdot)$. Each form $A_k(\cdot,\cdot)$ induces an operator $A_k : M_k \to M_k$ defined by

$$(A_k w, \varphi)_k = A_k(w, \varphi), \qquad \text{for all } \varphi \in M_k.$$

Denote by λ_k the largest eigenvalue of A_k.

These spaces are connected through $J-1$ linear operators $I_k : M_{k-1} \to M_k$, for $k = 2, \ldots, J$. These operators are often referred to as prolongation operators. The operators $Q_{k-1} : M_k \to M_{k-1}$ and $P_{k-1} : M_k \to M_{k-1}$ are defined by

$$(Q_{k-1} w, \varphi)_{k-1} = (w, I_k \varphi)_k, \qquad \text{for all } \varphi \in M_{k-1},$$

and

$$A_{k-1}(P_{k-1} w, \varphi) = A_k(w, I_k \varphi), \qquad \text{for all } \varphi \in M_{k-1}.$$

Finally, we assume that smoothers $R_k : M_k \to M_k$ are given and set

$$R_k^{(\ell)} = \begin{cases} R_k & \text{if } \ell \text{ is odd,} \\ R_k^t & \text{if } \ell \text{ is even.} \end{cases}$$

In the case discussed in Sections 2.1, $M_k \subset M_{k+1}$, $(\cdot,\cdot)_k = (\cdot,\cdot)$, $A_k(\cdot,\cdot) = A(\cdot,\cdot)$, $I_k = I$, and Q_k and P_k are projectors with respect to $(\cdot,\cdot)$ and $A(\cdot,\cdot)$ respectively. Here Q_k and P_k are not necessarily projectors. Note that the relationship $A_{k-1}P_{k-1} = Q_{k-1}A_k$ still holds.

4.2 General Multigrid Algorithms

Given $f \in M_J$, we are interested in solving

$$A_J u = f.$$

With u^0 given, we will consider the iterative algorithm

$$u^i = \mathrm{Mg}_J(u^{i-1}, f) \equiv u^{i-1} + B_J(f - A_J u^{i-1}),$$

where $B_J : M \to M$ is defined recursively by the following general multigrid procedure.

Algorithm 4.1 *Let p be a positive integer and let m_k be a positive integer depending on k. Set $B_1 = A_1^{-1}$. Assuming that $B_{k-1} : M_{k-1} \to M_{k-1}$ has been defined, we define $B_k : M_k \to M_k$ as follows. Let $g \in M_k$.*

(1) *Pre-smoothing: Set $x^0 = 0$ and define x^ℓ, $\ell = 1, \ldots, m_k$ by*

$$x^\ell = x^{\ell-1} + R_k^{(\ell+m_k)}(g - A_k x^{\ell-1}).$$

(2) *Correction: $y^{m_k} = x^{m_k} + I_k q^p$, where $q^0 = 0$ and q^i for $i = 1, \ldots, p$ is defined by*

$$q^i = q^{i-1} + B_{k-1}[Q_{k-1}(g - A_k x^{m_k}) - A_{k-1} q^{i-1}].$$

(3) *Post-smoothing: Define y^ℓ for $\ell = m_k + 1, \ldots, 2m_k$ by*

$$y^\ell = y^{\ell-1} + R_k^{(\ell+m_k)}(g - A_k y^{\ell-1}).$$

(4) $B_k g = y^{2m_k}$.

The cases $p = 1$ and $p = 2$ correspond to the V-cycle and the W-cycle multigrid algorithms respectively. The case $p = 1$ with the number of smoothings, m_k, varying is known as the variable V-cycle multigrid algorithm. In addition to the V-cycle and the W-cycle multigrid algorithms with the number of smoothings the same on each level, we will consider a variable V-cycle multigrid algorithm in which we will assume that the number of smoothings increases geometrically as k decreases. More precisely, in such a case we assume that there exist two constants β_0 and β_1 with $1 < \beta_0 \le \beta_1$ such that

$$\beta_0 m_k \le m_{k-1} \le \beta_1 m_k.$$

Note that the case $\beta_0 = \beta_1 = 2$ corresponds to doubling the number of smoothings as we proceed from M_k to M_{k-1}, i.e., $m_k = 2^{J-k} m_J$.

Our aim is to study the error reduction operator $\mathcal{E}_k = I - B_k A_k$ and provide conditions under which, we can estimate δ_k between zero and one such that

$$|A_k(\mathcal{E}_k u, u)| \le \delta_k A_k(u,u), \qquad \text{for all } u \in M_k.$$

We also show that the operator B_k corresponding to the variable V-cycle multigrid algorithm provides a good preconditioner for A_k even in the cases where $\mathcal{E}_k$ is not a reducer.

We first derive a recurrence relation. Let $K_k = I - R_k A_k$, $K_k^* = I - R_k^t A_k$ and set

$$K_k^{(m)} = \begin{cases} (K_k^* K_k)^{m/2} & \text{if } m \text{ is even} \\ (K_k^* K_k)^{\frac{m-1}{2}} K_k^* & \text{if } m \text{ is odd.} \end{cases}$$

By the definition of B_k the following two-level recurrence relation is easily derived:

$$\mathcal{E}_k = (K_k^{(m_k)})^* [I - I_k P_{k-1} + I_k \mathcal{E}_{k-1}^p P_{k-1}] K_k^{(m_k)}. \tag{33}$$

Our main assumption relating the spaces M_k and M_{k-1} is the following regularity-approximation property.

Condition (A.2") For some α with $0 < \alpha \le 1$ there exists C_{P} independent of k such that

$$|A_k((I - I_k P_{k-1})v, v)| \le C_{\mathrm{P}}^{2\alpha} \left(\frac{\|A_k v\|_k^2}{\lambda_k} \right)^{\alpha} [A_k(v,v)]^{1-\alpha}.$$

The so-called *variational assumption* refers to the case in which all of the forms $A_k(\cdot,\cdot)$ are defined in terms of the form $A_J(\cdot,\cdot)$ by

$$A_{k-1}(v,v) = A_k(I_k v, I_k v), \qquad \text{for all } v \in M_{k-1}. \tag{34}$$

We call this case the *inherited* case. Given the operators I_k, then all of forms come from $A(\cdot,\cdot)$. If the spaces are nested, then I_k can be chosen to be the natural injection operator and the forms can be defined by

$$A_k(v,v) = A(v,v), \qquad \text{for all } v \in M_k.$$

We call this case the *nested-inherited* case. It follows that in the nested-inherited case, Condition (A.2) implies Condition (A.2").

In general, (34) does not hold in the nonnested case. We will consider, at first, a weaker condition than (34).

Condition (I.1) For $k = 2, \ldots, J$, the operators I_k satisfy

$$A_k(I_k v, I_k v) \le A_{k-1}(v,v), \qquad \text{for all } v \in M_{k-1}.$$

It follows from the definition of P_{k-1} that Condition (I.1) holds if and only if

$$A_{k-1}(P_{k-1}v, P_{k-1}v) \leq A_k(v, v), \qquad \text{for all } v \in M_k$$

and if and only if

$$A_k((I - I_k P_{k-1})v, v) \geq 0, \qquad \text{for all } v \in M_k\,.$$

Lemma 4.1 *If Condition* (I.1) *holds then the error propagator* $\mathcal{E}_k$ *is symmetric and positive semidefinite with respect to* $A_k(\cdot,\cdot)$*, i.e.,*

$$A_k(\mathcal{E}_k v, v) = A_k((I - B_k A_k)v, v) \geq 0, \qquad \textit{for all } v \in M_k.$$

Proof. Since $A_k(I_k P_{k-1}w, v) = A_{k-1}(P_{k-1}w, P_{k-1}v)$, the operator $I_k P_{k-1}$ is symmetric with respect to $A_k(\cdot,\cdot)$. Hence by induction, since $\mathcal{E}_1 = I - B_1 A_1 = 0$, it follows, using (33), that $I - B_k A_k$ is symmetric and positive semidefinite. □

The assumption on the smoother is the same as in Section 2.1; we restate it here.

Condition (SM.1) There exists $\omega > 0$ not depending on J such that

$$\left(\frac{\omega}{\lambda_k}\right) \|v\|_k^2 \leq (\overline{R}_k v, v)_k, \qquad \text{for all } v \in M_k,\ k = 2, \ldots, J.$$

This is a condition local to M_k and hence does not depend on whether or not the spaces M_k are nested.

The following lemma is a generalization of Lemma 2.1.

Lemma 4.2 *If Condition* (SM.1) *holds, then*

$$\frac{\|A_k K_k^{(m)} v\|_k^2}{\lambda_k} \leq \frac{\omega^{-1}}{m}[A_k(v, v) - A_k(K_k^{(m)} v, K_k^{(m)} v)].$$

Proof. Let $\tilde{v} = K_k^{(m)} v$. By Condition (SM.1),

$$\frac{\|A_k \tilde{v}\|_k^2}{\lambda_k} \leq \omega^{-1} A_k(\overline{R}_k A_k \tilde{v}, \tilde{v}) = \omega^{-1} A_k((I - K_k^* K_k)\tilde{v}, \tilde{v}).$$

Suppose m is even. Then $\tilde{v} = (K_k^* K_k)^{m/2} v$. Set $\overline{K}_k = K_k^* K_k$. Then

$$\begin{aligned}\frac{\|A_k \tilde{v}\|_k^2}{\lambda_k} &\leq \omega^{-1} A_k((I - \overline{K}_k)\overline{K}_k^m v, v) \leq \frac{\omega^{-1}}{m} A_k((I - \overline{K}_k^m)v, v)\\ &= \frac{\omega^{-1}}{m}[A_k(v, v) - A_k(\tilde{v}, \tilde{v})].\end{aligned}$$

For m odd, we set $\overline{K}_k = K_k K_k^*$ and the result follows. □

4.3 Multigrid V-cycle as a Reducer

We will provide an estimate for the convergence rate of the V-cycle and the variable V-cycle multigrid algorithm under Conditions (A.2"), (I.1) and (SM.1).

It follows from Lemma 4.1 that if Condition (I.1) holds, then $A_k(\mathcal{E}_k u, u) \geq 0$. Hence our aim in such a case is to estimate δ_k between zero and one, where δ_k is such that

$$0 \leq A_k((I - B_k A_k)u, u) \leq \delta_k A_k(u, u), \qquad \text{for all } u \in M_k.$$

Theorem 4.1 (V-cycle) *Assume that Conditions* (A.2") *and* (I.1) *hold and that the smoothers satisfy Condition* (SM.1). *Let* $p = 1$. *Then*

$$0 \leq A_k(\mathcal{E}_k v, v) \leq \delta_k A_k(v, v), \qquad \textit{for all } v \in M_k \tag{35}$$

with $0 \leq \delta_k < 1$ *given in each case as follows:*

(a) *If* $p = 1$ *and* $m_k = m$ *for all* k, *then* (35) *holds with*

$$\delta_k = \frac{Mk^{\frac{1-\alpha}{\alpha}}}{Mk^{\frac{1-\alpha}{\alpha}} + m^\alpha} \quad \textit{with} \quad M = e^{(1-\alpha)}(\alpha C_{\mathrm{P}}^2/\omega + 1) - 1.$$

(b) *If* $p = 1$ *and* m_k *increases as* k *decreases, then* (35) *holds with*

$$\delta_k = 1 - \left(\frac{1}{1 + \frac{\alpha C_{\mathrm{P}}^2}{\omega m_k^\alpha}}\right) \prod_{i=2}^{k} \left(1 - \frac{1 - \alpha}{m_i^\alpha}\right).$$

In particular if there exist two constants β_0 *and* β_1 *with* $1 < \beta_0 \leq \beta_1$ *such that* $\beta_0 m_k \leq m_{k-1} \leq \beta_1 m_k$, *then* $\delta_k \leq \frac{M}{M + m_k^\alpha}$ *for* M *sufficiently large.*

Proof. By Lemma 4.1, the lower estimate in (35) follows from Condition (I.1). For the upper estimate, we claim that Conditions (A.2") and (SM.1) imply that

$$A_k(\mathcal{E}_k v, v) \leq \delta_k(\tau) A_k(v, v), \qquad \text{for all } v \in M_k \tag{36}$$

for $\tau > 0$ if $0 < \alpha < 1$ and for $\tau \geq \max_i(1/m_i^\alpha)$ if $\alpha = 1$. Here $\delta_k(\tau)$ is defined by

$$1 - \delta_k(\tau) = \frac{1}{1 + (\alpha C_{\mathrm{P}}^2/\omega)\tau} \prod_{i=2}^{k} \left(1 - \frac{(1-\alpha)}{(\tau m_i)^{\frac{\alpha}{1-\alpha}}}\right). \tag{37}$$

From this the theorem will follow by choosing an appropriate τ.

We shall prove (36) by induction. Since $\mathcal{E}_1 = I - B_1 A_1 = 0$, (36) holds for $k = 1$. Assume that

$$A_{k-1}(\mathcal{E}_{k-1} v, v) \leq \delta_{k-1}(\tau) A_{k-1}(v, v), \qquad \text{for all } v \in M_{k-1}.$$

Let $\tilde{v} = K_k^{(m_k)} v$. Using the two-level recurrence relation (33), we have

$$\begin{aligned}A_k(\mathcal{E}_k v, v) &\equiv A_k((I - B_k A_k)v, v)\\&= A_k((I - I_k P_{k-1})\tilde{v}, \tilde{v}) + A_{k-1}(\mathcal{E}_{k-1} P_{k-1}\tilde{v}, P_{k-1}\tilde{v})\\&\le A_k((I - I_k P_{k-1})\tilde{v}, \tilde{v}) + \delta_{k-1}(\tau) A_{k-1}(P_{k-1}\tilde{v}, P_{k-1}\tilde{v})\\&= A_k((I - I_k P_{k-1})\tilde{v}, \tilde{v}) + \delta_{k-1}(\tau) A_k(I_k P_{k-1}\tilde{v}, \tilde{v})\\&= [1 - \delta_{k-1}(\tau)]\, A_k((I - I_k P_{k-1})\tilde{v}, \tilde{v}) + \delta_{k-1}(\tau) A_k(\tilde{v}, \tilde{v}).\end{aligned}$$

By Conditions (A.2") and (SM.1) and using Lemma 4.2,

$$\begin{aligned}A_k((I - I_k P_{k-1})\tilde{v}, \tilde{v}) &\le C_{\mathrm{P}}^{2\alpha} \left(\frac{\|A_k \tilde{v}\|_k^2}{\lambda_k}\right)^{\alpha} [A_k(\tilde{v}, \tilde{v})]^{1-\alpha}\\&\le \frac{C_{\mathrm{P}}^{2\alpha}}{(\omega m_k)^{\alpha}} [A_k(v, v) - A_k(\tilde{v}, \tilde{v})]^{\alpha} [A_k(\tilde{v}, \tilde{v})]^{1-\alpha}.\end{aligned}$$

We now put things together to get

$$A_k(\mathcal{E}_k v, v) \le [1 - \delta_{k-1}(\tau)] \frac{C_{\mathrm{P}}^{2\alpha}}{(\omega m_k)^{\alpha}} (|||v|||^2 - |||\tilde{v}|||^2)^{\alpha} (|||\tilde{v}|||^2)^{1-\alpha} + \delta_{k-1}(\tau) |||\tilde{v}|||^2.$$

Let $x = |||\tilde{v}|||^2 / |||v|||^2$. Then

$$A_k(\mathcal{E}_k v, v) \le f(\delta_{k-1}(\tau); x, m_k) A_k(v, v),$$

where f is defined by

$$f(\delta; x, m) = (1 - \delta) \frac{C_{\mathrm{P}}^{2\alpha}}{(\omega m)^{\alpha}} (1 - x)^{\alpha} x^{1-\alpha} + \delta x, \qquad 0 \le x < 1. \tag{38}$$

By the Hölder inequality, we have

$$\frac{C_{\mathrm{P}}^{2\alpha}}{(\omega m)^{\alpha}} (1 - x)^{\alpha} x^{1-\alpha} \le (\alpha C_{\mathrm{P}}^2/\omega)\tau(1 - x) + (1 - \alpha)(\tau m)^{-\frac{\alpha}{1-\alpha}} x,$$

for all $\tau > 0$ if $0 < \alpha < 1$ and for all $\tau \ge 1/m^{\alpha}$ if $\alpha = 1$. Hence

$$f(\delta; x, m) \le \ell(\delta; x) \equiv \delta x + (1 - \delta)\Big[(\alpha C_{\mathrm{P}}^2/\omega)\tau(1 - x) + (1 - \alpha)(\tau m)^{-\frac{\alpha}{1-\alpha}} x\Big].$$

Note that $\ell(\delta; x)$ is linear in x and therefore

$$f(\delta; x, m) \le \ell(\delta; x) \le \max(\ell(\delta; 0), \ell(\delta; 1)).$$

If $\frac{(\alpha C_{\mathrm{P}}^2/\omega)\tau}{1 + (\alpha C_{\mathrm{P}}^2/\omega)\tau} \le \delta < 1$, then

$$\ell(\delta; 0) = (1 - \delta)(\alpha C_{\mathrm{P}}^2/\omega)\tau \le \delta \le \delta + (1 - \delta)\Big[(1 - \alpha)(\tau m)^{-\frac{\alpha}{1-\alpha}}\Big] = \ell(\delta; 1),$$

and hence

$$f(\delta; x, m) \leq \ell(\delta; 1) = \delta + (1-\delta)\Big[(1-\alpha)(\tau m)^{-\frac{\alpha}{1-\alpha}}\Big].$$

In particular, since $\frac{(\alpha C_{\mathrm{P}}^2/\omega)\tau}{1+(\alpha C_{\mathrm{P}}^2/\omega)\tau} = \delta_1(\tau) \leq \delta_{k-1}(\tau) < 1$,

$$f(\delta_{k-1}(\tau); x, m_k) \leq \delta_{k-1}(\tau) + [1-\delta_{k-1}(\tau)]\,[(1-\alpha)(\tau m_k)^{-\frac{\alpha}{1-\alpha}}] \equiv \delta_k(\tau).$$

As a consequence,

$$A_k(\mathcal{E}_k v, v) \leq f(\delta_{k-1}(\tau); x, m_k)\, A_k(v,v) \leq \delta_k(\tau)\, A_k(v,v)$$

holds for any $\tau > 0$ if $0 < \alpha < 1$ and for any $\tau \geq \max_i(1/m_i)$ if $\alpha = 1$. This proves (36).

If $m_k \equiv m$, then setting $\tau = k^{(1-\alpha)/\alpha}/m^\alpha$ in (37) shows that

$$\delta_k(\tau) \leq \frac{Mk^{\frac{1-\alpha}{\alpha}}}{Mk^{\frac{1-\alpha}{\alpha}} + m^\alpha}$$

with $M = e^{1-\alpha}(\alpha C_{\mathrm{P}}^2/\omega + 1) - 1 \leq C_{\mathrm{P}}^2/\omega + (e^{1-\alpha} - 1)$. This proves part (a) of the theorem.

Part (b) of the theorem follows by setting $\tau = m_k^{-\alpha}$ in (37) and noting that the product term is uniformly bounded from below if m_k increases geometrically as k decreases. □

Remark 1. Notice that, for $p = 1$ and $m_k = m$, $\delta_k \to 1$ as $k \to \infty$. This is in contrast to the results in the nested-inherited case under similar hypotheses. The deterioration, however is only like a power of k and hence may not be too serious.

Remark 2. For the variable V-cycle methods, if m_k increases geometrically as k decreases, then clearly the product term is strictly larger than 0 and thus δ_k is strictly less than 1, independently of k. So, e.g., if $m_J = 1$ we have

$$0 \leq A_J(\mathcal{E}_J v, v) \leq \frac{M}{M+1} A_J(v,v), \qquad \text{for all } v \in M_J.$$

If $m_k = 2^{J-k} m_J$, then the cost per iteration of the variable V-cycle multigrid algorithm is comparable to that of the W-cycle multigrid algorithm.

4.4 Multigrid W-cycle as a Reducer

We now provide an estimate for the rate of convergence of the W-cycle multigrid algorithm.

Theorem 4.2 (W-cycle) *Assume that Conditions* (A.2") *and* (I.1) *hold and that the smoothers satisfy Condition* (SM.1). *Let* $p = 2$ *and* $m_k = m$ *for all* k. *Then*

$$0 \le A_k(\mathcal{E}_k v, v) \le \delta A_k(v,v), \qquad \textit{for all } v \in M_k$$

and

$$\delta = \frac{M}{M+m^\alpha} \quad \textit{with } M \textit{ sufficiently large but independent of } k.$$

Proof. By Lemma 4.1, the lower estimate follows from Condition (I.1). We obtain the upper estimate inductively. Since $\mathcal{E}_1 = 0$, the estimate holds for $k = 1$. Assume that

$$A_{k-1}(\mathcal{E}_{k-1}v, v) \le \delta A_{k-1}(v,v), \qquad \text{for all } v \in M_{k-1}.$$

Then since $A_{k-1}(\mathcal{E}_{k-1}v, v) \ge 0$, we have

$$A_{k-1}(\mathcal{E}_{k-1}^2 v, v) \le \delta^2 A_{k-1}(v,v), \qquad \text{for all } v \in M_{k-1}.$$

Let $\tilde{v} = K_k^{(m)} v$. Using the two-level recurrence (33), we obtain for $p = 2$,

$$\begin{aligned} A_k(\mathcal{E}_k v, v) &\equiv A_k((I - B_k A_k)v, v) \\ &= A_k((I - I_k P_{k-1})\tilde{v}, \tilde{v}) + A_{k-1}(\mathcal{E}_{k-1}^2 P_{k-1}\tilde{v}, P_{k-1}\tilde{v}) \\ &\le A_k((I - I_k P_{k-1})\tilde{v}, \tilde{v}) + \delta^2 A_{k-1}(P_{k-1}\tilde{v}, P_{k-1}\tilde{v}) \\ &= A_k((I - I_k P_{k-1})\tilde{v}, \tilde{v}) + \delta^2 A_k(I_k P_{k-1}\tilde{v}, \tilde{v}) \\ &= (1-\delta^2) A_k((I - I_k P_{k-1})\tilde{v}, \tilde{v}) + \delta^2 A_k(\tilde{v}, \tilde{v}). \end{aligned}$$

By Conditions (A.2") and (SM.1), and using Lemma 4.2,

$$\begin{aligned} A_k((I - I_k P_{k-1})\tilde{v}, \tilde{v}) &\le C_{\mathrm{P}}^{2\alpha} \left(\frac{\|A_k \tilde{v}\|_k^2}{\lambda_k} \right)^\alpha [A_k(\tilde{v}, \tilde{v})]^{1-\alpha} \\ &\le \frac{C_{\mathrm{P}}^{2\alpha}}{(\omega m)^\alpha} [A_k(v,v) - A_k(\tilde{v}, \tilde{v})]^\alpha [A_k(\tilde{v}, \tilde{v})]^{1-\alpha}. \end{aligned}$$

We now put things together to get

$$A_k(\mathcal{E}_k v, v) \le (1-\delta^2) \frac{C_{\mathrm{P}}^{2\alpha}}{(\omega m)^\alpha} (|||v|||^2 - |||\tilde{v}|||^2)^\alpha (|||\tilde{v}|||^2)^{1-\alpha} + \delta^2 |||\tilde{v}|||^2.$$

Let $x = |||\tilde{v}|||^2 / |||v|||^2$. Then

$$A_k(\mathcal{E}_k v, v) \le f(\delta^2; x, m) A_k(v,v),$$

with $f(\delta; x, m)$ defined by (38). The theorem will follow if we can show

$$f(\delta^2; x, m) \le \delta, \tag{39}$$

for $\delta = M/(M + m^\alpha)$ with M sufficient large.

We now prove (39). We have shown in the proof of Theorem 4.1 that

$$f(\delta^2; x, m) \le \max(\ell(\delta^2; 0), \ell(\delta^2; 1)), \qquad \text{for all } \tau > 0.$$

Here $\ell(\delta; x) \equiv \delta x + (1-\delta)[(\alpha C_{\mathrm{P}}^2/\omega)\tau(1-x) + (1-\alpha)(\tau m)^{-\frac{\alpha}{1-\alpha}} x]$. It thus suffices to show that there exists a number δ with $0 < \delta < 1$ such that, with an appropriately chosen τ (with the possible restrictions $\tau \ge 1/m$ for $\alpha = 1$),

$$\ell(\delta^2; 0) = (1-\delta^2)(\alpha C_{\mathrm{P}}^2/\omega)\, \tau \le \delta$$

and

$$\ell(\delta^2; 1) = \delta^2 + (1-\delta^2)(1-\alpha)(\tau m)^{-\alpha/(1-\alpha)} \le \delta.$$

These two inequalities can be written as

$$\left(\frac{1+\delta}{\delta}\right)^{1-\alpha} \left(\frac{1-\alpha}{m^{\alpha/(1-\alpha)}}\right)^{1-\alpha} \le \tau^\alpha \le \left(\frac{\delta}{1-\delta^2}\right)^{\alpha} \left(\frac{1}{\alpha C_{\mathrm{P}}^2/\omega}\right)^{\alpha}.$$

This is equivalent to choosing a $\delta \in (0,1)$ so that

$$\frac{1+\delta}{\delta}(1-\delta)^\alpha \le \frac{m^\alpha}{(\alpha C_{\mathrm{P}}^2/\omega)^\alpha (1-\alpha)^{1-\alpha}} \equiv \frac{m^\alpha}{C_\alpha}$$

holds uniformly in $m \ge 1$. Clearly, we can take $\delta = \frac{M}{M+m^\alpha}$ with M large enough such that above inequality holds. For example, take $M \ge \frac{1}{2}(4C_\alpha)^{1/\alpha} = \frac{1}{2}4^{1/\alpha}(\alpha C_{\mathrm{P}}^2/\omega)(1-\alpha)^{\frac{(1-\alpha)}{\alpha}}$. We have thus proved (39) and hence the theorem. □

In many applications, Condition (I.1) is not valid. We note that Condition (I.1) is used in Lemma 4.1 to prove

$$A_k(\mathcal{E}_k v, v) \equiv A_k((I - B_k A_k)v, v) \ge 0.$$

Without (I.1), we have to prove

$$|A_k(\mathcal{E}_k v, v)| \le \delta A_k(v, v) \quad \text{for some } \delta < 1.$$

It is sufficient to assume either the number of smoothings, m, to be sufficiently large (but independent of k), or the following stability condition on I_k.

Condition (I.2) For $k = 2, \ldots, J$, the operators I_k satisfy

$$A_k(I_k v, I_k v) \le 2A_{k-1}(v, v), \qquad \text{for all } v \in M_{k-1}.$$

Theorem 4.3 (W-cycle) *Suppose that Condition* (A.2") *holds and that the smoothers, R_k, satisfy Condition* (SM.1). *Let $p = 2$ and $m_k = m$ for all k. Assume that either*

(a) *Condition* (I.2) *holds or*
(b) *the number of smoothings, m, is sufficiently large.*

Then

$$|A_k(\mathcal{E}_k v, v)| \le \delta A_k(v,v), \qquad \textit{for all } v \in M_k, \tag{40}$$

with $\delta = \frac{M}{M+m^\alpha}$ *with* M *sufficiently large.*

Proof. We proceed by induction. Let δ be defined as in Theorem 4.2. Since $\mathcal{E}_k = I - B_1 A_1 = 0$, (40) holds for $k = 1$. Assume that

$$|A_{k-1}(\mathcal{E}_{k-1} v, v)| \le \delta A_{k-1}(v,v), \qquad \text{for all } v \in M_{k-1}.$$

Then

$$0 \le A_{k-1}(\mathcal{E}_{k-1}^2 v, v) \le \delta^2 A_{k-1}(v,v), \qquad \text{for all } v \in M_{k-1}.$$

We can prove in a way similar to the proof of the previous theorem that

$$A_k(\mathcal{E}_k v, v) \le \delta A_k(v,v),$$

with $\delta = M/(M + m^\alpha)$ given by the previous theorem. We now show that

$$-A_k(\mathcal{E}_k v, v) \le \delta A_k(v,v),$$

with the same δ. We first consider the case in which m is sufficiently large. Note that $\mathcal{E}_k^2$ is always symmetric, positive semidefinite. The two-level recurrence relation (33), with $p = 2$, implies that

$$-A_k(\mathcal{E}_k v, v) \le -A_k((I - I_k P_{k-1})\tilde{v}, \tilde{v}) \le \frac{C_{\mathrm{P}}^{2\alpha}}{(\omega m)^\alpha} A_k(v,v).$$

Without loss of generality, we assume that $M > C_{\mathrm{P}}^{2\alpha}/\omega^\alpha$. Then we choose m so large that $\frac{C_{\mathrm{P}}^{2\alpha}}{(\omega m)^\alpha} \le \delta = \frac{M}{M+m^\alpha}$ and then (40) follows in this case.

Now if Condition (I.2) holds, then

$$-A_k((I - I_k P_{k-1})\tilde{v}, \tilde{v}) \le A_k(\tilde{v}, \tilde{v}).$$

This implies that for any $\delta \in (0,1)$,

$$-A_k(\mathcal{E}_k v, v) \le (1-\delta^2)|A_k((I - I_k P_{k-1})\tilde{v}, \tilde{v})| + \delta^2 A_k(\tilde{v}, \tilde{v}).$$

This is the same bound as for $A_k(\mathcal{E}_k v, v)$. Thus the proof proceeds as before to show that

$$|A_k(\mathcal{E}_k v, v)| \le \delta A_k(v,v),$$

for $\delta = \frac{M}{M+m^\alpha}$ with M sufficiently large. This proves the theorem. □

4.5 Multigrid V-cycle as a Preconditioner

We now consider using the multigrid operator B_k as a preconditioner for A_k. Any symmetric positive definite operator can be used as a preconditioner for A_k. We will discuss the conditions under which multigrid operator B_k is a good preconditioner.

Theorem 4.1 states that if Conditions (A.2") and (I.1) hold and the smoothers satisfy Condition (SM.1), then

$$0 \leq A_k(\mathcal{E}_k v, v) \equiv A_k((I - B_k A_k)v, v) \leq \delta_k A_k(v, v)$$

which is equivalent to

$$(1 - \delta_k) A_k(v, v) \leq A_k(B_k A_k v, v) \leq A_k(v, v). \tag{41}$$

Therefore, in such a case, B_k is an optimal preconditioner for A_k.

In many situations, Condition (I.1) is not satisfied. In such a case $A_k(\mathcal{E}_k v, v)$ could become negative, and thus $\mathcal{E}_k$ may not be a reducer. For the V-cycle and the variable V-cycle multigrid algorithms, we have proved without using Condition (I.1) that $A_k(\mathcal{E}_k v, v) \leq \delta_k A_k(v, v)$, which is equivalent to the lower estimate in (41). Condition (I.1) is only used to get the upper estimate in (41). We do not need the upper estimate in (41) in order to use B_k as a preconditioner. For a good preconditioner, we want to find numbers $\underline{\eta}_k$ and $\overline{\eta}_k$ such that

$$\underline{\eta}_k A_k(v, v) \leq A_k(B_k A_k v, v) \leq \overline{\eta}_k A_k(v, v), \qquad \text{for all } v \in M_k. \tag{42}$$

We will provide estimates for $\underline{\eta}_k$ and $\overline{\eta}_k$ for the variable V-cycle multigrid operator under the Conditions (SM.1) and (A.2"). Note again that (SM.1) and (A.2") imply the lower estimate in (41), and hence the lower estimate of (42) follows with $\underline{\eta}_k \geq 1 - \delta_k$.

Theorem 4.4 (V-cycle preconditioner) *Let* $p = 1$. *Assume that Condition* (A.2") *holds and that the smoothers,* R_k, *satisfy Condition* (SM.1). *Then* (42) *holds with*

$$\underline{\eta}_k = \frac{\omega m_k^\alpha}{\alpha C_{\mathrm{P}}^2 + \omega m_k^\alpha} \prod_{i=2}^{k} \left(1 - \frac{1-\alpha}{m_i^\alpha}\right) \quad \textit{and} \quad \overline{\eta}_k = \prod_{i=2}^{k} (1 + \bar{\delta}_i)$$

where $\bar{\delta}_i = \frac{C_{\mathrm{P}}^{2\alpha}}{\omega^\alpha m_i^\alpha} \alpha^\alpha (1-\alpha)^{1-\alpha}$. *In particular if the number of smoothings,* m_k, *satisfies*

$$\beta_0 m_k \leq m_{k-1} \leq \beta_1 m_k$$

for some $1 < \beta_0 \leq \beta_1$ *independent of* k, *then*

$$\underline{\eta}_k \geq \frac{m_k^\alpha}{M + m_k^\alpha} \quad \textit{and} \quad \overline{\eta}_k \leq 1 + \frac{M}{m_k^\alpha} = \frac{M + m_k^\alpha}{m_k^\alpha} \quad \textit{for some } M.$$

Proof. We have shown in the proof of Theorem 4.1 that Conditions (A.2") and (SM.1) imply that $A_k((I - B_k A_k)v, v) = A_k(\mathcal{E}_k v, v) \le \delta_k A_k(v, v)$. Consequently, $(1 - \delta_k) A_k(v, v) \le A_k(B_k A_k v, v)$ and the lower estimate in (42) holds with

$$\underline{\eta}_k = 1 - \delta_k = \left(\frac{1}{\alpha C_{\mathrm{P}}^2/(\omega m_k^\alpha) + 1} \right) \prod_{i=2}^{k} \left(1 - \frac{1-\alpha}{m_i^\alpha} \right).$$

We prove the upper estimate in (42) by induction. For $k = 1$ there is nothing to prove. Assume (42) is true for $k - 1$. Then

$$-A_{k-1}((I - B_{k-1}A_{k-1})w, w) \le (\overline{\eta}_{k-1} - 1) A_{k-1}(w, w), \qquad \text{for all } w \in M_{k-1}.$$

Set $\tilde{v} = K_k^{(m_k)} v$. By the induction hypothesis

$$\begin{aligned}
-A_k((I &- B_k A_k)v, v) \\
&= -A_k((I - I_k P_{k-1})\tilde{v}, \tilde{v}) - A_{k-1}((I - B_{k-1}A_{k-1})P_{k-1}\tilde{v}, P_{k-1}\tilde{v}) \\
&\le -A_k((I - I_k P_{k-1})\tilde{v}, \tilde{v}) + (\overline{\eta}_{k-1} - 1) A_{k-1}(P_{k-1}\tilde{v}, P_{k-1}\tilde{v}) \\
&= -\overline{\eta}_{k-1} A_k((I - I_k P_{k-1})\tilde{v}, \tilde{v}) + (\overline{\eta}_{k-1} - 1) A_k(\tilde{v}, \tilde{v}) \\
&\le -\overline{\eta}_{k-1} A_k((I - I_k P_{k-1})\tilde{v}, \tilde{v}) + (\overline{\eta}_{k-1} - 1) A_k(v, v).
\end{aligned} \tag{43}$$

It remains to estimate $-A_k((I - I_k P_{k-1})\tilde{v}, \tilde{v})$. By Condition (A.2")

$$-A_k((I - I_k P_{k-1})\tilde{v}, \tilde{v}) \le C_{\mathrm{P}}^{2\alpha} \left[\frac{\|A_k \tilde{v}\|_k^2}{\lambda_k} \right]^\alpha [A_k(\tilde{v}, \tilde{v})]^{1-\alpha}.$$

By Lemma 4.2,

$$\frac{\|A_k \tilde{v}\|_k^2}{\lambda_k} \le \frac{1}{\omega m_k} [A_k(v, v) - A_k(\tilde{v}, \tilde{v})].$$

Hence, since $0 \le A_k(\tilde{v}, \tilde{v}) \le A_k(v, v)$, we have that

$$\begin{aligned}
-A_k((I &- I_k P_{k-1})\tilde{v}, \tilde{v}) \\
&\le \frac{C_{\mathrm{P}}^{2\alpha}}{(\omega m_k)^\alpha} [A_k(v, v) - A_k(\tilde{v}, \tilde{v})]^\alpha [A_k(\tilde{v}, \tilde{v})]^{1-\alpha} \\
&\le \frac{C_{\mathrm{P}}^{2\alpha}}{(\omega m_k)^\alpha} \alpha^\alpha (1 - \alpha)^{1-\alpha} A_k(v, v) \\
&= \bar{\delta}_i A_k(v, v).
\end{aligned} \tag{44}$$

Combining (43) and (44), we obtain

$$\begin{aligned}
-A_k((I &- B_k A_k)v, v) \\
&\le [\overline{\eta}_{k-1} \bar{\delta}_k + (\overline{\eta}_{k-1} - 1)] A_k(v, v) \\
&= (\overline{\eta}_k - 1) A_k(v, v).
\end{aligned}$$

Hence $A_k(B_k A_k v, v) \le \overline{\eta}_k A_k(v, v)$ and the theorem is proved. □

We now prove a general result for the V-cycle multigrid without assuming the regularity-approximation assumption (A.2").

Theorem 4.5 *If the operators $\overline{R}_k$, $k = 1, 2, \ldots, J$, are symmetric and positive definite, then the operator B_J corresponding to the V-cycle multigrid method ($p = 1$) is symmetric and positive definite. If in addition, Condition (SM.1) holds, then*

$$\frac{\omega}{\lambda_k} A_k(v, v) \leq A_k((B_k A_k)v, v).$$

Proof. It is the same to prove that $B_k A_k$ is symmetric and positive definite with respect to $A_k(\cdot, \cdot)$. The symmetry is easy to see. Let $\tilde{v} = K_k^{(m_k)} v$. Since $p = 1$, we get

$$A_k((I - B_k A_k)v, v) = A_k(\tilde{v}, \tilde{v}) - A_{k-1}((B_{k-1} A_{k-1}) P_{k-1} \tilde{v}, P_{k-1} \tilde{v}).$$

The induction hypothesis implies that

$$A_k((I - B_k A_k)v, v) < A_k(\tilde{v}, \tilde{v}).$$

If $\overline{R}_k$ is symmetric and positive definite, then $A_k(\tilde{v}, \tilde{v}) < A_k(v, v)$ and B_k is symmetric and positive definite. If in addition, (SM.1) holds, then $A_k(\tilde{v}, \tilde{v}) \leq (1 - \omega/\lambda_k) A_k(v, v)$ and thus $\omega \lambda_k^{-1} A_k(v, v) \leq A_k(B_k A_k v, v)$. □

The theorem shows that the (variable) V-cycle multigrid operator B_k is symmetric and positive definite, and thus can always be used as a preconditioner if the size of the condition number is not a concern. If in addition we know that Condition (I.1) holds, then by Lemma 4.1, we have

$$0 \leq A_k((I - B_k A_k)v, v), \qquad \text{for all } v \in M_k$$

or equivalently

$$A_k(B_k A_k v, v) \leq A_k(v, v), \qquad \text{for all } v \in M_k.$$

Therefore, if Conditions (I.1) and (SM.1) hold, then

$$A_k(\overline{R}_k A_k v, v) \leq A_k(B_k A_k v, v) \leq A_k(v, v).$$

5 Computational Scales of Sobolev Norms

5.1 Introduction

In this lecture we provide a framework for developing computationally efficient multilevel preconditioners and representations for Sobolev norms. The material, for the most part, is taken from [6]. Specifically, given a Hilbert space V and a nested sequence of subspaces, $V_1 \subset V_2 \subset \ldots \subset V$, we construct operators which are spectrally equivalent to those of the form $\mathcal{A} = \sum_k \mu_k (Q_k - Q_{k-1})$. Here μ_k, $k = 1, 2, \ldots$ are positive numbers and

Q_k is the orthogonal projector onto V_k with $Q_0 = 0$. We first present abstract results which show when $\mathcal{A}$ is spectrally equivalent to a similarly constructed operator $\widetilde{\mathcal{A}}$ defined in terms of an approximation $\widetilde{Q}_k$ of Q_k , for $k = 1, 2, \ldots$.

We describe how these results lead to efficient preconditioners for discretizations of differential and pseudo-differential operators of positive and negative order and to sums of such operators.

Multilevel subspace decompositions provide tools for the construction of preconditioners. One of the first examples of such a construction was provided in [9] where a simple additive multilevel operator (BPX) was developed for preconditioning second order elliptic boundary value problems. The analysis of the BPX preconditioner involves the verification of norm equivalences of the form

$$\|u\|^2_{H^1(\Omega)} \simeq \sum_{k=1}^{J} h_k^{-2} \|(Q_k - Q_{k-1})u\|^2_{L^2(\Omega)}, \qquad \text{for all } u \in V_J. \tag{45}$$

The above norms are those corresponding to the Sobolev space $H^1(\Omega)$ and $L^2(\Omega)$ respectively. The quantity h_k is the approximation parameter associated with V_k. The original results in [9] were sharpened by [23] and [30] to show that (45) holds with constants of equivalence independent of J. Practical preconditioners involve the replacement of the operator $Q_k - Q_{k-1}$ by easily computable operators as discussed in [9].

In addition to the above application, there are other practical applications of multilevel decompositions. In particular, for boundary element methods, it is important to have computationally simple operators which are equivalent to pseudo-differential operators of order one and minus one. In addition, multilevel decompositions which provide norm equivalences for $H^{1/2}(\partial\Omega)$ can be used to construct bounded extension operators used in nonoverlapping domain decomposition with inexact subdomain solves.

The equivalence (45) is the starting point of the multilevel analysis. This equivalence is valid for $J = \infty$ in which case we get a norm equivalence on $H^1(\Omega)$. It follows from (45) that

$$\|v\|^2_{H^s(\Omega)} \simeq \sum_{k=1}^{\infty} h_k^{-2s} \|(Q_k - Q_{k-1})v\|^2_{L^2(\Omega)},$$

for $s \in [0, 1]$. Here $\|\cdot\|_{H^s(\Omega)}$ denotes the norm on the Sobolev space $H^s(\Omega)$ of order s. This means that the operator

$$\mathcal{A}^s = \sum_{k=1}^{\infty} h_k^{-2s} (Q_k - Q_{k-1}) \tag{46}$$

can be used as a preconditioner. However, $\mathcal{A}^s$ is somewhat expensive to evaluate since the evaluation of the projector Q_k requires the solution of a Gram matrix problem. Thus, many researchers have sought computationally efficient operators which are equivalent to $\mathcal{A}^s$.

Some techniques for constructing such operators based on wavelet or wavelet–like space decompositions are given by [13], [18], [20], [26], [25], [28], [29] and others. In the domain decomposition literature, extension operators that exploit multilevel decomposition were used in [10], [16], and [22].

In this lecture we construct simple multilevel decomposition operators which can also be used to define norms equivalent to the usual norms on Sobolev spaces. Specifically, we develop computationally efficient operators which are uniformly equivalent to the more general operator

$$\mathcal{A}_J = \sum_{k=1}^{J} \mu_k (Q_k - Q_{k-1}), \tag{47}$$

where $1 \leq J \leq \infty$ and $\{\mu_k\}$ are positive constants. We start by stating an abstract theorem. The proof is in [6]. Let $\{\widetilde{Q}_k\}$, with $\widetilde{Q}_k : V_J \to V_k$, be another sequence of linear operators. The theorem shows that the operators $\mathcal{A}_J$ and

$$\widetilde{\mathcal{A}}_J = \sum_{k=1}^{J} \mu_k (\widetilde{Q}_k^t - \widetilde{Q}_{k-1}^t)(\widetilde{Q}_k - \widetilde{Q}_{k-1}) \tag{48}$$

are spectrally equivalent under appropriate assumptions on the spaces V_k, the operators $\widetilde{Q}_k$ and the sequence $\{\mu_k\}$. Here $\widetilde{Q}_k^t$ is the adjoint of $\widetilde{Q}_k$. The abstract results are subsequently applied to develop efficient preconditioners when $\widetilde{Q}_k$ is defined in terms of a simple averaging operator. Some partial results involving the operator used here were stated in [22].

Because of the generality of the abstract results, they can be applied to preconditioning sums of operators. An example of this is the so-called "singularly perturbed" problem resulting from preconditioning parabolic time stepping problems which leads to

$$\mu_k = (\epsilon h_k^{-2} + 1)^{-1}.$$

In this example ϵ is the time step size. Our results give rise to preconditioned systems with uniformly bounded condition numbers independent of the parameter ϵ.

We note that [25] provides an L^2-stable local basis for the spaces $\{\text{Range}(Q_k - Q_{k-1})\}$. With such a construction it is possible to obtain preconditioners for many of the applications considered in this lecture. However, our approach is somewhat simpler to implement. In addition, our abstract framework allows for easy application to other situations such as function spaces which are piecewise polynomials of higher order.

5.2 A Norm Equivalence Theorem

We now provide abstract conditions which imply the spectral equivalence of (47) and (48). We start by introducing the multilevel spaces. Let V be a

Hilbert space with inner product $(\cdot,\cdot)$. We assume that we are given a nested sequence of approximation subspaces,

$$V_1 \subset V_2 \subset \ldots \subset V,$$

and that this sequence is dense in V. Let θ_j, $j = 1, 2, \ldots$, be a non-decreasing sequence of positive real numbers. Define H to be the subspace of V such that the norm

$$|||v||| = \left(\sum_{j=1}^{\infty} \theta_j \|(Q_j - Q_{j-1})v\|^2 \right)^{1/2}$$

is finite. Here $\|\cdot\|$ denotes the norm in V, Q_j for $j > 0$, denotes the orthogonal projection onto V_j and $Q_0 = 0$. Clearly H is a Hilbert space and $\{V_k\}$ is dense in H.

The following properties are obvious from the construction.

1. The "inverse inequality" holds for V_j, i.e.,
$$|||v||| \leq \theta_j^{1/2} \|v\|, \qquad \text{for all } v \in V_j. \tag{49}$$
2. The "approximation property" holds for V_j, i.e.,
$$\|(Q_j - Q_{j-1})v\| \leq \theta_j^{-1/2} |||v|||, \qquad \text{for all } v \in H. \tag{50}$$

As discussed in the introduction, the abstract results will be stated in terms of an additional sequence of "approximation" operators, $\widetilde{Q}_k : V \to V_k$ for $k > 0$ and $\widetilde{Q}_0 = 0$. These operators are assumed to satisfy the following three conditions, for $k = 1, 2, \ldots$.

1. An "approximation property": There exists a constant C_A such that
$$\|(Q_k - \widetilde{Q}_k)v\| \leq C_A \theta_k^{-1/2} |||v|||, \qquad \text{for all } v \in H. \tag{51}$$
2. Uniform coercivity of $\widetilde{Q}_k$: There exists a $\delta > 0$ such that
$$\delta \|v_k\|^2 \leq (\widetilde{Q}_k v_k, v_k), \qquad \text{for all } v_k \in V_k. \tag{52}$$
3. The range of $\widetilde{Q}_k^t$, the adjoint of $\widetilde{Q}_k$, is contained in V_k. This condition is equivalent to
$$\widetilde{Q}_k Q_k = Q_k \widetilde{Q}_k. \tag{53}$$

Remark 5.1 *Let $\{\phi_i\}_{i=1}^m$ be a basis for V_k. It is not difficult to see that there exists $\{f_i\}_{i=1}^m$ with $f_i \in V$ such that*

$$\tilde{Q}_k v = \sum_{i=1}^{m} (v, f_i)\ \phi_i \qquad \textit{for all } v \in V.$$

Then

$$\tilde{Q}_k^t w = \sum_{i=1}^{m} (w, \phi_i)\ f_i \qquad \textit{for all } w \in V_k.$$

Thus Condition 3 above holds if and only if $f_i \in V_k$, for $i = 1, \ldots, m$.

The purpose of this section is to provide abstract conditions which guarantee that the symmetric operators $\mathcal{A}_J$ and $\widetilde{\mathcal{A}}_J$, defined respectively by (47) and (48), are spectrally equivalent. Let $\mathcal{L} = (\ell_{k,j})$ be the lower triangular (infinite) matrix with nonzero entries

$$\ell_{k,j} = \left(\frac{\theta_j \mu_k}{\theta_k \mu_j}\right)^{1/2}, \ k \geq j. \tag{54}$$

We assume that $\mathcal{L}$ has bounded l_2 norm, i.e.,

$$\|\mathcal{L}\|_{\ell_2} \equiv \sup_{\{\xi_k\},\,\{\zeta_k\}} \frac{\sum_{k=1}^{\infty} \sum_{j\leq k} \ell_{k,j}\, \xi_k \zeta_j}{\left(\sum_{k=1}^{\infty} \xi_k^2\right)^{1/2} \left(\sum_{k=1}^{\infty} \zeta_k^2\right)^{1/2}} \leq C_{\mathcal{L}}. \tag{55}$$

The above condition implies that

$$\mu_k \leq C\theta_k$$

for $C = C_{\mathcal{L}}^2 \mu_1/\theta_1$. Thus, $(\mathcal{A}_J v, v) < \infty$ for all $v \in H$.

We introduce one final condition: There exists a constant α such that

$$\mu_k + \mu_{k+1} \leq \alpha \mu_k, \ \text{for } k = 1, 2, \ldots. \tag{56}$$

We can now state the main abstract theorem.

Theorem 5.1 *Assume that conditions (51)–(53), (55), and (56) are satisfied. Then the operator $\widetilde{\mathcal{A}}_J$ defined by (48), with $1 \leq J \leq \infty$, satisfies*

$$[3(1+\alpha\delta^{-2}C_A^2 C_{\mathcal{L}}^2)]^{-1}(\mathcal{A}_J v, v) \leq (\widetilde{\mathcal{A}}_J v, v) \leq 3(1+\alpha C_A^2 C_{\mathcal{L}}^2)\,(\mathcal{A}_J v, v),$$

for all $v \in H$.

Remark 5.2 *If W is the completion of H under the norm $\|v\|_{\mathcal{A}} = (\mathcal{A}_\infty v, v)^{1/2}$, then the estimate of Theorem 5.1 extends to all of W by density.*

With the following lemma the theorem is easily proved. Its proof is in [6].

Lemma 5.1 *Assume that conditions (51)–(53), and (55) are satisfied. Then for all $u \in H$,*

$$\sum_{k=1}^{J} \mu_k \|(Q_k - \widetilde{Q}_k)u\|^2 \leq C_A^2 C_{\mathcal{L}}^2\,(\mathcal{A}_J u, u) \tag{57}$$

and

$$\sum_{k=1}^{J} \mu_k \|(Q_k - \widetilde{Q}_k)u\|^2 \leq \delta^{-2} C_A^2 C_{\mathcal{L}}^2\,(\widetilde{\mathcal{A}}_J u, u). \tag{58}$$

Proof (Proof of Theorem 5.1). Note that

$$(\widetilde{Q}_k - \widetilde{Q}_{k-1}) = (Q_k - Q_{k-1}) - (Q_k - \widetilde{Q}_k) + (Q_{k-1} - \widetilde{Q}_{k-1}).$$

Thus for $v \in H$,

$$\begin{aligned}(\widetilde{\mathcal{A}}_J v, v) &= \sum_{k=1}^{J} \mu_k \|(\widetilde{Q}_k - \widetilde{Q}_{k-1})v\|^2 \\ &\leq 3\Big(\sum_{k=1}^{J} \mu_k \|(Q_k - Q_{k-1})v\|^2 + \sum_{k=1}^{J} (\mu_k + \mu_{k+1})\|(Q_k - \widetilde{Q}_k)v\|^2\Big) \\ &\leq 3(1 + \alpha C_A^2 C_{\mathcal{L}}^2)\,(\mathcal{A}_J v, v).\end{aligned}$$

We used (56) and Lemma 5.1 for the last inequality above. The proof for the other inequality is essentially the same. This completes the proof of the theorem. □

5.3 Development of Preconditioners

The above results can be applied to the development of preconditioners. Indeed, consider preconditioning an operator on V_J which is spectrally equivalent to

$$L_J = \sum_{k=1}^{J} \mu_k^{-1} (Q_k - Q_{k-1}). \tag{59}$$

Our preconditioner B_J is to be spectrally equivalent to the operator

$$A_J \equiv L_J^{-1} = \sum_{k=1}^{J} \mu_k (Q_k - Q_{k-1}).$$

Let

$$B_J = \sum_{k=1}^{J} \mu_k (\widetilde{Q}_k - \widetilde{Q}_{k-1})^t (\widetilde{Q}_k - \widetilde{Q}_{k-1}). \tag{60}$$

Then B_J and A_J are spectrally equivalent provided that $\{\mu_k\}$ and $\{\widetilde{Q}_k\}$ satisfy the hypothesis of the theorem. It follows that $B_J L_J$ is well conditioned.

5.4 Preconditioning Sums of Operators

We next consider the case of preconditioning sums of operators. Suppose $\{\hat{\mu}_k\}$ is another sequence which satisfies conditions (55) and (56). Then

$$\hat{L}_J = \sum_{k=1}^{J} \hat{\mu}_k^{-1} (Q_k - Q_{k-1}). \tag{61}$$

can be preconditioned by the operator defined by replacing μ_k by $\hat{\mu}_k$ in (60) above. The following corollary shows that the result can be extended to non-negative combinations of L_J and $\hat{L}_J$.

Corollary 5.1 *Assume that conditions (51)–(53) are satisfied and that (55) and (56) hold for both $\{\mu_k\}$ and $\{\hat{\mu}_k\}$. For nonnegative c_1, c_2 with $c_1 + c_2 > 0$ define*

$$B_J = \sum_{k=1}^{J} (c_1\mu_k^{-1} + c_2\hat{\mu}_k^{-1})^{-1} (\widetilde{Q}_k - \widetilde{Q}_{k-1})^t (\widetilde{Q}_k - \widetilde{Q}_{k-1}). \tag{62}$$

Then for $1 \le J \le \infty$,

$$[3(1 + 4\alpha\delta^{-2}C_A^2C_{\mathcal{L}}^2)]^{-1}((c_1L_J + c_2\hat{L}_J)^{-1}v, v) \le (B_Jv, v) \le 3(1 + 4\alpha C_A^2C_{\mathcal{L}}^2)\,((c_1L_J + c_2\hat{L}_J)^{-1}v, v), \quad \textit{for all } v \in H.$$

The above corollary shows that B_J is spectrally equivalent to $(c_1L_J + c_2\hat{L}_J)^{-1}$ and hence provides a uniform preconditioner for $c_1L_J + c_2\hat{L}_J$. Moreover, the resulting condition number (for the preconditioned system) is bounded independently of the parameters c_1 and c_2.

Proof. Note that

$$(c_1L_J + c_2\hat{L}_J)^{-1} = \sum_{k=1}^{J} (c_1\mu_k^{-1} + c_2\hat{\mu}_k^{-1})^{-1} (Q_k - Q_{k-1}).$$

To apply the theorem to this operator, we simply must check the conditions on the sequence $\widetilde{\mu}_k = (c_1\mu_k^{-1} + c_2\hat{\mu}_k^{-1})^{-1}$. The corresponding lower triangular matrix has entries

$$(\widetilde{\mathcal{L}})_{k,j} = \left(\frac{\theta_j\widetilde{\mu}_k}{\theta_k\widetilde{\mu}_j}\right)^{1/2} = \left(\frac{\theta_j(c_1\mu_j^{-1} + c_2\hat{\mu}_j^{-1})}{\theta_k(c_1\mu_k^{-1} + c_2\hat{\mu}_k^{-1})}\right)^{1/2}$$
$$\le \left(\frac{\theta_j}{\theta_k}\left(\frac{\mu_k}{\mu_j} + \frac{\hat{\mu}_k}{\hat{\mu}_j}\right)\right)^{1/2} \le \left(\frac{\theta_j\mu_k}{\theta_k\mu_j}\right)^{1/2} + \left(\frac{\theta_j\hat{\mu}_k}{\theta_k\hat{\mu}_j}\right)^{1/2} = (\mathcal{L} + \hat{\mathcal{L}})_{k,j}.$$

Since $0 \le (\widetilde{\mathcal{L}})_{k,j} \le (\mathcal{L} + \hat{\mathcal{L}})_{k,j}$, for every pair k, j, it follows that

$$\|\widetilde{\mathcal{L}}\|_{\ell_2} \le \|\mathcal{L} + \hat{\mathcal{L}}\|_{\ell_2} \le 2C_{\mathcal{L}}.$$

Because (56) holds for both $\{\mu_k\}$ and $\{\hat{\mu}_k\}$, it clearly holds for $\{\widetilde{\mu}_k\}$. The corollary follows by application of the theorem. □

5.5 A Simple Approximation Operator $\widetilde{Q}_k$

In this section, we define and analyze a simple approximation operator $\widetilde{Q}_k$. Our applications involve Sobolev spaces with possibly mixed boundary conditions.

Let Ω be a polygonal domain in R^2 with boundary $\partial\Omega = \Gamma_D \cup \Gamma_N$ where Γ_D and Γ_N are essentially disjoint. Dirichlet boundary conditions are imposed on Γ_D. We consider domains in R^2 for convenience. Generalizations of the results to be presented to domains in R^d, with $d > 2$, at least for rectangular parallelepipeds, are straightforward.

For non-negative integers s, let $H^s(\Omega)$ denote the Sobolev space of order s on Ω (see, e.g. [14],[15]). The corresponding norm and semi-norm are denoted $\|\cdot\|_{H^s(\Omega)}$ and $|\cdot|_{H^s(\Omega)}$ respectively. The space $H^1_D(\Omega)$ is defined to be the functions in $H^1(\Omega)$ which vanish on Γ_D and for $s > 1$, $H^s_D(\Omega) = H^s(\Omega) \cap H^1_D(\Omega)$. For positive non-integers s, the spaces $H^s(\Omega)$ and $H^s_D(\Omega)$ are defined by interpolation between the neighboring integers using the real method of Lions and Peetre (cf. [15]). For negative s, $H^s(\Omega)$ is defined to be the space of linear functionals for which the norm

$$\|u\|_{H^s(\Omega)} = \sup_{\phi \in H_D^{-s}(\Omega)} \frac{< u, \phi >}{\|\phi\|_{H_D^{-s}(\Omega)}}$$

is finite. Here $< \cdot, \cdot >$ denotes the duality pairing. Clearly, for $s < 0$, $L^2(\Omega) \subseteq H^s(\Omega)$ if we identify $u \in L^2(\Omega)$ with the functional $< u, \phi > \equiv (u, \phi)$.

Some Basic Approximation Properties

Let $\mathcal{T}$ be a locally quasi-uniform triangulation of Ω and τ be a closed triangle in $\mathcal{T}$ with diameter h_τ. Let $\widetilde{\tau}$ be the subset of the triangles in $\mathcal{T}$ whose boundaries intersect τ and define $V_{\widetilde{\tau}}$ to be the finite element approximation subspace consisting of functions which are continuous on $\widetilde{\tau}$ and piecewise linear with respect to the triangles of $\widetilde{\tau}$. Note that there are no boundary conditions imposed on the elements of $V_{\widetilde{\tau}}$. We restrict the discussion in this paper to piecewise linear subspaces. Extensions of these considerations to more general nodal finite element subspaces pose no significant additional difficulties.

The following facts are well known.

1. Given $u \in H^1(\widetilde{\tau})$, there exists a constant $\widetilde{u}$ such that

$$\|u - \widetilde{u}\|_{H^s(\widetilde{\tau})} \le C h_\tau^{1-s} |u|_{H^1(\widetilde{\tau})}, \quad s = 0, 1. \tag{63}$$

2. Given $u \in H^2(\widetilde{\tau})$, there exists a linear function $\widetilde{u}$ such that

$$\|u - \widetilde{u}\|_{H^s(\widetilde{\tau})} \le C h_\tau^{2-s} |u|_{H^2(\widetilde{\tau})}, \quad s = 0, 1, 2. \tag{64}$$

The best constants satisfying the above inequalities clearly depend on the shape of the domain $\widetilde{\tau}$. However, under the assumption that the triangulation is locally quasi-uniform, it is possible to show that the above inequalities hold with constants only depending on s and on the quasi-uniformity constants.

For the purpose of analyzing our multilevel example we define the following local approximation operator $\widetilde{Q}_{\widetilde{\tau}} : L^2(\Omega) \to V_{\widetilde{\tau}}$. Let ϕ_i, $i = 1, 2, \ldots, m$, be the nodal basis for $V_{\widetilde{\tau}}$. The operator $\widetilde{Q}_{\widetilde{\tau}}$ is given by

$$\widetilde{Q}_{\widetilde{\tau}} u = \sum_{i=1}^{m} \frac{(u, \phi_i)_{\widetilde{\tau}}}{(1, \phi_i)_{\widetilde{\tau}}} \phi_i, \tag{65}$$

with $(\cdot,\cdot)_{\widetilde{\tau}}$ the inner product in $L^2(\widetilde{\tau})$. For $u, v \in L^2(\widetilde{\tau})$,

$$(\widetilde{Q}_{\widetilde{\tau}} u, v)_{\widetilde{\tau}} = \sum_{i=1}^{m} \frac{(u, \phi_i)_{\widetilde{\tau}} (v, \phi_i)_{\widetilde{\tau}}}{(1, \phi_i)_{\widetilde{\tau}}}$$

and hence it immediately follows that $\widetilde{Q}_{\widetilde{\tau}}$ is symmetric on $L^2(\widetilde{\tau})$. Moreover, $\widetilde{Q}_{\widetilde{\tau}}$ is positive definite when restricted to $V_{\widetilde{\tau}}$ (see, Lemma 5.5). The next lemma provides a basic approximation property for $\widetilde{Q}_{\widetilde{\tau}}$.

Lemma 5.2 *Let τ be in $\mathcal{T}$. Then for $s = 0, 1$, there exists a constant C, independent of τ, such that*

$$\|u - \widetilde{Q}_{\widetilde{\tau}} u\|_{L^2(\widetilde{\tau})} \leq C h_\tau^s \|u\|_{H^s(\widetilde{\tau})}, \qquad \textit{for all } u \in H^s(\widetilde{\tau}). \tag{66}$$

Proof. A simple computation shows that

$$\|\widetilde{Q}_{\widetilde{\tau}} u\|_{L^2(\widetilde{\tau})} \leq C \|u\|_{L^2(\widetilde{\tau})}$$

from which (66) immediately follows for $s = 0$. For $s = 1$, let $\tilde{u}$ be the constant function satisfying (63). Using the previous estimate, since $\widetilde{Q}_{\widetilde{\tau}} \tilde{u} = \tilde{u}$, we have

$$\|u - \widetilde{Q}_{\widetilde{\tau}} u\|_{L^2(\widetilde{\tau})} \leq \|u - \tilde{u}\|_{L^2(\widetilde{\tau})} + \|\widetilde{Q}_{\widetilde{\tau}}(u - \tilde{u})\|_{L^2(\widetilde{\tau})} \leq C \|u - \tilde{u}\|_{L^2(\widetilde{\tau})}.$$

Combining the above inequalities and (63) completes the proof of (66) for $s = 1$. □

Approximation Properties: the Multilevel Case

We provide some stronger approximation properties in the case when the mesh results from a multilevel refinement strategy. Again we describe the case of $d = 2$. The analogous constructions for $d > 2$, at least for the case of rectangular parallelepipeds, are straightforward generalizations. Assume that an initial coarse triangulation $\mathcal{T}_1$ of Ω has been provided with Γ_D aligning with the mesh $\mathcal{T}_1$. By this we mean that any edge of $\mathcal{T}_1$ on $\partial\Omega$ is either contained in Γ_D or intersects Γ_D at most at the endpoints of the edge. Multilevel triangulations are defined recursively. For $k > 1$, the triangulation $\mathcal{T}_k$ is defined by breaking each triangle in $\mathcal{T}_{k-1}$ into four, by connecting the centers of the edges. The finite element space V_k consists of the functions which are continuous on Ω, piecewise linear with respect to $\mathcal{T}_k$ and vanish on Γ_D. Let $h_k = \max_{\tau \in \mathcal{T}_k} h_\tau$. Clearly, $h_k = 2^{-k+1} h_1$.

We now define a sequence of approximation operators $\widetilde{Q}_k : L^2(\Omega) \to V_k$. Let ϕ_i, $i = 1, \ldots, m$ be the nodal basis for V_k. We define $\widetilde{Q}_k$ by

$$\widetilde{Q}_k u = \sum_{i=1}^{m} \frac{(u, \phi_i)}{(1, \phi_i)} \phi_i. \tag{67}$$

Remark 5.3 *Let τ be a triangle of $\mathcal{T}_k$. It is easy to see that $\widetilde{Q}_{\tilde{\tau}}u$ and $\widetilde{Q}_k u$ agree on τ as long as $\tau \cap \Gamma_D = \emptyset$.*

In the multilevel case, we have the following stronger version of Lemma 5.2.

Lemma 5.3 *Let s be in $[0, 3/2)$. There exists a constant C_s not depending on h_k such that*

$$\|u - \widetilde{Q}_k u\|_{L^2(\Omega)} \le C_s h_k^s \|u\|_{H^s(\Omega)}, \qquad \textit{for all } u \in H_D^s(\Omega).$$

For the proof of the lemma, we will use the following lemma which is a slight modification of Lemma 6.1 of [8]. Its proof is contained in the proof of Lemma 6.1 of [8].

Lemma 5.4 *Let Ω^η denote the strip $\{x \in \Omega \,|\, dist(x, \partial\Omega) < \eta\}$ and $0 \le s < 1/2$. Then for all $v \in H^{1+s}(\Omega)$,*

$$\|v\|_{H^1(\Omega^\eta)} \le C\eta^s \|v\|_{H^{1+s}(\Omega)}. \tag{68}$$

In addition, let Ω_D^η denote the strip $\{x \in \Omega \,|\, dist(x, \Gamma_D) < \eta\}$. Then for all v in $H_D^1(\Omega)$,

$$\|v\|_{L^2(\Omega_D^\eta)} \le C\eta \, \|v\|_{H^1(\Omega^\eta)}. \tag{69}$$

Proof (Proof of Lemma 5.3). The proof for $s = 0$ is trivial (see, Lemma 5.2). For positive s, we consider two cases. First we examine triangles whose boundaries do not intersect the boundary of any triangle in $\mathcal{T}_1$. We shall denote this set by $\tau \cap \mathcal{T}_1 = \emptyset$ and the remaining set of triangles by $\tau \cap \mathcal{T}_1 \ne \emptyset$.

Let ϕ_i be the nodal basis function in the space V_k associated with the node x_i^k. Assume that x_i^k does not lie on the boundary of any triangle $\tau \in \mathcal{T}_1$. Because of the multilevel construction, the mesh $\mathcal{T}_k$ is symmetric with respect to reflection through the point x_i^k. It follows that the nodal basis function ϕ_i, restricted to a line passing through x_i^k, is an even function with respect to x_i^k. Let $x_i^k = (p_1, p_2)$. Then, both of the functions $x - p_1$ and $y - p_2$ are odd on each such line. Consequently,

$$(x - p_1, \phi_i) = (y - p_2, \phi_i) = 0.$$

Thus, it follows from Remark 5.3 that $\widetilde{Q}_k \widetilde{u}(x_i^k) = \widetilde{u}(x_i^k)$ for any linear function $\widetilde{u}$.

Let τ be a triangle whose boundary does not intersect the boundary of any triangle of $\mathcal{T}_1$. Applying the above argument to each node of τ shows that $\widetilde{Q}_k \widetilde{u} = \widetilde{u}$ on τ for any linear function $\widetilde{u}$. Let $\tilde{\tau}$ be as in Lemma 5.2. Given $u \in H^2(\tilde{\tau})$, let $\tilde{u}$ be the linear function satisfying (64). As in the proof of Lemma 5.2, we get

$$\|u - \widetilde{Q}_k u\|_{L^2(\tau)} = \|u - \widetilde{Q}_{\tilde{\tau}} u\|_{L^2(\tau)} \le C\|u - \tilde{u}\|_{L^2(\tilde{\tau})} \le C h_k^s \|u\|_{H^s(\tilde{\tau})}$$

for $s = 0, 1, 2$. Summing the above inequality and interpolating gives

$$\left(\sum_{\tau \cap \mathcal{T}_1 = \emptyset} \|u - \widetilde{Q}_k u\|_{L^2(\tau)}^2 \right)^{1/2} \leq C h_k^s \|u\|_{H^s(\Omega)} \tag{70}$$

for $s \in [0, 2]$.

We next consider the case when τ intersects an edge in the triangulation $\mathcal{T}_1$. Suppose that τ intersects Γ_D. We clearly have that

$$\|\widetilde{Q}_k u\|_{L^2(\tau)} \leq C \|u\|_{L^2(\widetilde{\tau})}. \tag{71}$$

Thus,

$$\|u - \widetilde{Q}_k u\|_{L^2(\tau)} \leq C \|u\|_{L^2(\widetilde{\tau})}.$$

Summing the above inequality and applying (69) gives

$$\left(\sum_{\tau \cap \Gamma_D \neq \emptyset} \|u - \widetilde{Q}_k u\|_{L^2(\tau)}^2 \right)^{1/2} \leq C h_k \|u\|_{H^1(\Omega^{2h_k})}. \tag{72}$$

Finally, we consider the case when τ intersects an edge in the triangulation $\mathcal{T}_1$ and does not intersect Γ_D. By Remark 5.3 and Lemma 5.2,

$$\|u - \widetilde{Q}_k u\|_{L^2(\tau)} \leq C h_k \|u\|_{H^1(\widetilde{\tau})}.$$

Summing the above inequality and using (72) gives

$$\left(\sum_{\tau \cap \mathcal{T}_1 \neq \emptyset} \|u - \widetilde{Q}_k u\|_{L^2(\tau)}^2 \right)^{1/2} \leq C h_k \|u\|_{H^1(E^{2h_k})}. \tag{73}$$

Here E^{2h_k} denotes the strip of width $\mathcal{O}(2h_k)$ around all element edges from the initial triangulation $\mathcal{T}_1$.

The lemma for $s = 1$ follows combining (70) and (73). The result for $s \in (0, 1)$ follows by interpolation. For $1 < s < 3/2$, (68) and (73) imply

$$\left(\sum_{\tau \cap \mathcal{T}_1 \neq \emptyset} \|u - \widetilde{Q}_k u\|_{L^2(\tau)}^2 \right)^{1/2} \leq C h_k^s \|u\|_{H^s(\Omega)}.$$

The lemma for $1 < s < 3/2$ follows combining the above inequality with (70). This completes the proof of the lemma. □

Remark 5.4 *We can extend these arguments to the case when V_k consists of piecewise quadratic functions with respect to the k'th triangulation. Again $\{\phi_k\}$ denotes the nodal basis for V_k. Then $\tilde{Q}_k$ defined by (67) satisfies Lemma 5.3. The proof is identical to the case of linears.*

The Coercivity Estimate

We next show that the coercivity estimate (52) holds for $\widetilde{Q}_k$. Actually, we only require that the triangulation $\mathcal{T}_h$ be locally quasi-uniform. We assume that Γ_D aligns with this triangulation and let V_h be the functions which are piecewise linear with respect to this triangulation, continuous on Ω and vanish on Γ_D. We consider the linear operator $\widetilde{Q}_h$ defined analogously to $\widetilde{Q}_k$ in (67) and show that

$$\|v\|^2 \le C(\widetilde{Q}_h v, v), \qquad \text{for all } v \in V_h.$$

The constant C above only depends on the quasi-uniformity constant (or minimal angle).

Let $\{x_i\}$ for $i = 1, \dots, m$ be the nodes of the triangulation and $\{\phi_i\}$ be the corresponding nodal basis functions. The mesh is quasi-uniform so for each ϕ_i, there is a parameter h_i such that

$$h_\tau \simeq h_i \tag{74}$$

holds for all triangles τ which have the node x_i as a vertex. Here we define $a \simeq b$ to mean that

$$a \le Cb \text{ and } b \le Ca$$

with constant C independent of the triangulation. It is well known that

$$(v, v) \simeq \sum_{\tau \in \mathcal{T}_h} h_\tau^2 \sum_{x_l \in \tau} v(x_l)^2, \qquad \text{for all } v \in V_h. \tag{75}$$

It follows from (74) that

$$(v, v) \simeq \sum_{i=1}^{m} h_i^2\, v(x_i)^2, \qquad \text{for all } v \in V_h. \tag{76}$$

We can now prove the coercivity estimate. This result was essentially given in [9] for the case of a globally quasi-uniform triangulation.

Lemma 5.5 *Assume that the mesh $\mathcal{T}_h$ is locally quasi-uniform. There is a constant C only depending on the quasi-uniformity condition such that*

$$C^{-1}(v, v) \le (\widetilde{Q}_h v, v) \le C\ (v, v), \qquad \textit{for all } v \in V_h.$$

Proof. Let G be the Gram matrix, i.e.

$$G_{ij} = (\phi_i, \phi_j), \quad i, j = 1, \dots, m$$

and D be the diagonal matrix with entries $D_{ii} = h_i^2$. Let v be in V_h and w be the coefficient vector satisfying

$$v = \sum_{i=1}^{m} w_i \phi_i.$$

Note that (76) can be rewritten as

$$C^{-1}((Gw,w)) \le ((Dw,w)) \le C((Gw,w)), \qquad \text{for all } w \in R^m.$$

Here $((\cdot,\cdot))$ denotes the inner product on R^m. This is equivalent to

$$C^{-1}((D^{-1}Gw,Gw)) \le ((Gw,w)) \le C((D^{-1}Gw,Gw)), \qquad \text{for all } w \in R^m.$$

Since

$$(1,\phi_i) \simeq h_i^2,$$

it follows that

$$\begin{aligned}(\widetilde{Q}_h v, v) &= \sum_{i=1}^m \frac{(v,\phi_i)^2}{(1,\phi_i)} = \sum_{i=1}^m \frac{((Gw)_i)^2}{(1,\phi_i)} \\ &\simeq ((D^{-1}Gw,Gw)) \simeq ((Gw,w)) = (v,v).\end{aligned}$$

This completes the proof of the lemma. □

5.6 Applications

Here we apply some of the above results. We follow [6]. As we have seen in the previously, the operator $\widetilde{Q}_k$ satisfies the approximation and coercivity estimates required for application of the abstract results. Throughout this discussion we assume that $V_1 \subset V_2 \subset \ldots$ is a sequence of nested piecewise linear and continuous multilevel spaces as described earlier. We take $V = L^2(\Omega)$ and $(\cdot,\cdot)$ to be the corresponding inner product. With a slight abuse of notation we also use $(\cdot,\cdot)$ to denote the obvious duality pairing.

Remark 5.5 *Since $V_k \subset H^s(\Omega)$, for $0 \le s < 3/2$, Q_k and $\widetilde{Q}_k$ extend naturally to all of $H^{-s}(\Omega)$. Let $-3/2 < s < 3/2$ and define $\mathcal{A}^s$ as in (46). It is known that the norm $(\mathcal{A}^s u, u)^{1/2}$ is equivalent to $\|\cdot\|_{H^s(\Omega)}$; cf. [24].*

Fix $\gamma < 3/2$. By Lemma 5.3, the triangle inequality and well known properties of Q_k

$$\|(Q_k - \widetilde{Q}_k)u\|_{L^2(\Omega)} \le C\theta_k^{-1/2}\|u\|_{H^\gamma(\Omega)}$$

where $\theta_k = h_k^{-2\gamma}$. Let $s < \gamma$ and set $\mu_k = h_k^{-2s}$. Then,

$$\ell_{k,j} = \left(\frac{h_k}{h_j}\right)^{\gamma - s}$$

decays exponentially as a function of $k-j$. An elementary computation gives that

$$\|\mathcal{L}\| \le C_{\mathcal{L}} = \left(1 - \left(\frac{1}{2}\right)^{\gamma-s}\right)^{-1}.$$

The next theorem immediately follows from Remark 5.5, Remark 5.2 and Theorem 5.1.

Theorem 5.2 *Let $-3/2 < s < 3/2$. Then $(\widetilde{\mathcal{A}}^{(s)}u,u)^{1/2}$ provides a norm on $H^s_D(\Omega)$ which is equivalent to the usual Sobolev norm. Here*

$$\widetilde{\mathcal{A}}^{(s)}u = \sum_{k=1}^{\infty} h_k^{-2s}(\widetilde{Q}_k - \widetilde{Q}_{k-1})^2 u.$$

A Preconditioning Example

We consider applying the earlier results to develop a preconditioner for an example involving a pseudo-differential operator of order minus one. The canonical example of such an application is associated with a form

$$\mathcal{V}(u,v) = \int_\Omega \int_\Omega \frac{u(s_1)v(s_2)}{|s_1 - s_2|}\, ds_1 ds_2.$$

For this application, Γ_D is empty and we seek preconditioners for the problem: Find $U \in V_J$ satisfying

$$\mathcal{V}(U,\phi) = F(\phi) \quad \text{for all } \phi \in V_J.$$

Here F is a given functional. It is shown in [5] that

$$\mathcal{V}(u,u) \simeq \|u\|^2_{H^{-1/2}(\Omega)} \quad \text{for all } u \in V_J. \tag{77}$$

It is convenient to consider the problem of preconditioning in terms of operators. Specifically, let $\mathcal{V} : V_J \to V_J$ be defined by

$$(\mathcal{V}v,w) = \mathcal{V}(v,w) \quad \text{for all } v,w \in V_J.$$

We shall see that $\widetilde{\mathcal{A}}_J^{(1/2)}$ defined by

$$\widetilde{\mathcal{A}}_J^{(1/2)} = \sum_{k=1}^{J} h_k^{-1}(\widetilde{Q}_k - \widetilde{Q}_{k-1})^2$$

provides a computationally efficient preconditioner for $\mathcal{V}$. Indeed, by Theorem 5.1,

$$(\widetilde{\mathcal{A}}_J^{(1/2)}u,u) \simeq (\mathcal{A}^{1/2}u,u) \quad \text{for all } u \in V_J.$$

Applying Remark 5.5 and (77) implies that

$$(\mathcal{V}\mathcal{A}^{1/2}u, \mathcal{A}^{1/2}u) \simeq (\mathcal{A}^{-1/2}\mathcal{A}^{1/2}u, \mathcal{A}^{1/2}u) = (u, \mathcal{A}^{1/2}u)$$

for all $u \in V_J$. Thus, $\widetilde{\mathcal{A}}_J^{(1/2)}\mathcal{V}$ has a bounded spectral condition number.

It is easy to evaluate the action of $\widetilde{\mathcal{A}}_J^{(1/2)}$ in a preconditioned iteration procedure. For $k = 1,2,\ldots,J$, let $\{\phi_i^k\}$ denote the nodal basis for V_k. In typical preconditioning applications, one is required to evaluate the action of the preconditioner on a function v where only the quantities $\{(v,\phi_i^J)\}$ are known. One could, of course, compute v from $\{(v, \phi_i^J)\}$ but this would require solving a Gram matrix problem. Our preconditioner avoids the Gram matrix problem. To evaluate the action of $\widetilde{Q}_k$, for $1 \le k \le J$, one is only required to take linear combinations of the quantities $\{(v, \phi_i^k)\}$. Note that (v, ϕ_i^k) is a simple linear combination of $\{(v, \phi_i^{k+1})\}$. Thus, we see that all of the $\widetilde{Q}_k$'s can be computed efficiently (with work proportional to the number of unknowns on the finest level J) by a V-cycle-like algorithm.

Two Examples Involving Sums of Operators

The first example involves preconditioning the discrete systems which result from time stepping a parabolic initial value problem. For the second example we consider a Tikhonov regularization of a problem with noisy data.

Fully discrete time stepping schemes for parabolic problems often lead to problems of the form: Find $u \in S_h$ satisfying

$$(u, \phi) + \epsilon D(u, \phi) = F(\phi) \quad \text{for all } \phi \in S_h. \tag{78}$$

Here $D(\cdot, \cdot)$ denotes the Dirichlet form on Ω and S_h is the finite element approximation. The parameter ϵ is related to the time step size and is often small. Assume that $S_h = V_J$ where V_J is a multilevel approximation space as developed earlier. Let $\mu_k = 1$ and $\hat{\mu}_k = h_k^2$, for $k = 1, 2, \ldots$. For convenience, we assume that Γ_D is non-empty so that $D(v, v) \simeq \|v\|_1^2$, for all $v \in H_D^1(\Omega)$. Then for L_J and $\hat{L}_J$ defined respectively by (59) and (61), we have

$$(L_J v, v) \simeq (v, v) \text{ and } (\hat{L}_J v, v) \simeq D(v, v)$$

for all $v \in V_J$. Applying Corollary 5.1 gives that

$$B_J = \sum_{k=1}^{J} (\mu_k^{-1} + \epsilon \hat{\mu}_k^{-1})^{-1} (\widetilde{Q}_k - \widetilde{Q}_{k-1})^2 \tag{79}$$

provides a uniform preconditioner for the discrete operator associated with (78). The resulting condition number for the preconditioned system can be bounded independently of the time step size ϵ and the number of levels J.

We next consider an example which results from Tikhonov regularization of a problem with noisy data. We consider approximating the solution of the problem

$$Tv = f$$

where T denotes the inverse of the Laplacian and $f \in L^2(\Omega)$. This is replaced by the discrete problem

$$T_h v = f_h$$

where T_h is the Galerkin solution operator, i.e., $T_h v = w$ where $w \in V_J$ satisfies

$$D(w, \theta) = (v, \theta) \quad \text{for all } \theta \in V_J$$

and f_h is the $L^2(\Omega)$ orthogonal projection onto V_J. If it is known that v is smooth but f is noisy, better approximations result from regularization [21], [27]. We consider the regularized solution $\widetilde{w} \in V_J$ satisfying

$$(T_h + \alpha A_h)\widetilde{w} = f_h. \tag{80}$$

Here $A_h : V_J \to V_J$ is defined by

$$(A_h v, w) = D(v, w) \quad \text{for all } v, w \in V_J.$$

The regularization parameter α is often small (see, [27]) and can be chosen optimally in terms of the magnitude of the noise in f.

Preconditioners for the sum in (80) of the form of (79) result from the application of Corollary 5.1. In this case, $\mu_k = h_k^{-2}$, $\hat{\mu}_k = h_k^2$. The condition numbers for the resulting preconditioned systems can be bounded independent of the regularization parameter α.

Preconditioners for systems like (80) are generally not easily developed. The problem is that the operator applied to the higher frequencies (depending on the size of α) behaves like a differential operator while on the lower frequencies, it behaves like the inverse of a differential operator. This causes difficulty in most multilevel methods.

$H^1(\Omega)$ Bounded Extensions

We finally consider the construction of $H^1(\Omega)$ bounded extensions. Such extensions are useful in development of domain decomposition preconditioners with inexact subdomain solves. The construction given here is essentially the same as that in [22]. We include it here in detail as an application of Theorem 4.1.

With $\{V_j\}$ as above, let $\widetilde{V}_k$ (for $k = 1, 2, \ldots, J$) be the functions defined on $\partial\Omega$ which are restrictions of those in V_k. This gives a multilevel structure on the finest space $\widetilde{V}_J$. These spaces inherit a nodal basis from the original nodal basis on V_k. The nodal basis function associated with a boundary node x_i is just the restriction of the basis function for V_k associated with x_i. Denoting this basis by $\{\psi_i^k\}$, we define

$$\widetilde{q}_k(f) = \sum \frac{< f, \psi_i^k >}{< 1, \psi_i^k >} \psi_i^k.$$

The above sum is taken over the nodal basis elements for $\widetilde{V}_k$ and $< \cdot, \cdot >$ denotes the $L^2(\partial\Omega)$ inner product. We note that it is known [24] that

$$\|\theta\|^2_{H^{1/2}(\partial\Omega)} \simeq \sum_{k=1}^{J} h_k^{-1} \|(q_k - q_{k-1})\theta\|^2_{L^2(\partial\Omega)}$$

where q_k denotes the L^2-projection onto $\widetilde{V}_k$. It is easy to see that Theorem 4.1 holds for these spaces. Thus

$$\|\theta\|^2_{H^{1/2}(\partial\Omega)} \simeq \sum_{k=1}^{J} h_k^{-1} \|(\widetilde{q}_k - \widetilde{q}_{k-1})\theta\|^2_{L^2(\partial\Omega)}, \tag{81}$$

with $\widetilde{q}_J\theta = \theta$ and $\widetilde{q}_0\theta = 0$.

Now given a function $\theta \in \widetilde{V}_J$, we define $E_J\theta \in V_J$ by $E_J\theta = \sum_{k=1}^{J} \omega_k$ with ω_k defined as follows. Let $\bar{\theta}$ be the mean value of θ on $\partial\Omega$. Then ω_1 is the function in V_1 defined by

$$\omega_1(x_i) = \widetilde{q}_1(x_i), \qquad \text{if } x_i \text{ is a node of } V_1 \text{ on } \partial\Omega,$$

and

$$\omega_1(x_i) = \bar{\theta}, \qquad \text{if } x_i \text{ is a node of } V_1 \text{ in the interior of } \Omega.$$

For $J \geq k > 1$, ω_k is the function in V_k defined by

$$\omega_k(x_i) = [\widetilde{q}_k\theta - \widetilde{q}_{k-1}\theta](x_i), \qquad \text{if } x_i \text{ is a node of } V_k \text{ on } \partial\Omega,$$

and

$$\omega_k(x_i) = 0, \qquad \text{if } x_i \text{ is a node of } V_k \text{ in the interior of } \Omega.$$

Note that $E_J\theta = \theta$ on $\partial\Omega$ so that E_J is an extension operator.

Recall that $|\cdot|_{H^1(\Omega)}$ denotes the semi-norm on $H^1(\Omega)$. Then

$$|E_J\theta|_{H^1(\Omega)} = |E_J\theta - \bar{\theta}|_{H^1(\Omega)} = |E_J(\theta - \bar{\theta})|_{H^1(\Omega)} \leq \|E_J(\theta - \bar{\theta})\|_{H^1(\Omega)}.$$

We now use the following well known multilevel characterization of the $H^1(\Omega)$ norm on V_J:

$$\|v\|^2_{H^1(\Omega)} \simeq \inf \sum_{k=1}^{J} h_k^{-2} \|v_k\|^2_{L^2(\Omega)},$$

where the infimum is taken over all splittings $v = \sum_{k=1}^{J} v_k$, with $v_k \in V_k$. Applying this with $v = E_J(\theta - \bar{\theta}) = (\omega_1 - \bar{\theta}) + \sum_{k=2}^{J} \omega_k$ and using (81), we conclude that

$$\begin{aligned}
\|E_J(\theta - \bar{\theta})\|^2_{H^1(\Omega)} &\leq C\left[\sum_{k=2}^{J} h_k^{-2} \|\omega_k\|^2_{L^2(\Omega)} + h_1^{-2} \|\omega_1 - \bar{\theta}\|^2_{L^2(\Omega)}\right] \\
&\leq C \sum_{k=1}^{J} h_k^{-1} \|(\widetilde{q}_k - \widetilde{q}_{k-1})(\theta - \bar{\theta})\|^2_{L^2(\partial\Omega)} \\
&\leq C\|\theta - \bar{\theta}\|^2_{H^{1/2}(\partial\Omega)} \leq C|\theta|^2_{H^{1/2}(\partial\Omega)},
\end{aligned}$$

where $|\cdot|_{H^{1/2}(\partial\Omega)}$ denotes the $H^{1/2}(\partial\Omega)$ semi-norm. Thus we see that

$$|E_J\theta|_{H^1(\Omega)} \leq C|\theta|_{H^{1/2}(\partial\Omega)}.$$

This type of bounded extension operator is precisely what is required for the development of non-overlapping domain decomposition algorithms with inexact solves.

References

1. R.A. Adams. *Sobolev Spaces.* Academic Press, Inc., New York, 1975.
2. D. Braess and W. Hackbusch. A new convergence proof for the multigrid method including the V-cycle. *SIAM J. Numer. Anal.*, 20:967–975, 1983.
3. J. H. Bramble and X. Zhang. The analysis of a multigrid methods. In *Handbook of Numerical Analysis, Vol. VII.* North-Holland, Amsterdam, 2001.
4. James H. Bramble. On the development of multigrid methods and their analysis. In *Proceedings of Symposia in Applied Mathematics*, volume 48, pages 5–19, 1994.
5. James H. Bramble, Zbigniew Leyk, and Joseph E. Pasciak. Iterative schemes for nonsymmetric and indefinite elliptic boundary value problems. *Math. Comp.*, 60:1–22, 1993.
6. James H. Bramble, J. E. Pasciak, and P. Vasilevski. Computational scales of sobolev spaces with application to preconditioning. *Math. Comp.*, 69:463–480, 2001.
7. James H. Bramble and Joseph E. Pasciak. New convergence estimates for multigrid algorithms. *Math. Comp.*, 49:311–329, 1987.
8. James H. Bramble and Joseph E. Pasciak. New estimates for multigrid algorithms including the V-cycle. *Math. Comp.*, 60:447–471, 1993.
9. James H. Bramble, Joseph E. Pasciak, and Jinchao Xu. Parallel multilevel preconditioners. *Math. Comp.*, 55:1–22, 1990.
10. James H. Bramble and Panayot S. Vassilevski. Wavelet–like extension operators in interface domain decomposition methods. 1997, 1998.
11. James H. Bramble and J. Xu. Some estimates for weighted L^2 projections. *Math. Comp.*, 56:463–476, 1991.
12. P.L. Butzer and H. Berens. *Semi-Groups of Operators and Approximation.* Springer-Verlag, New York, 1967.
13. J. M. Carnicer, Wolfgang Dahmen, , and J. M. Peña. Local decompositions of refinable spaces and wavelets. *Appl. Comp. Harm. Anal.*, 3:127–153, 1996.
14. P. G. Ciarlet. *The Finite Element Method for Elliptic Problems.* North-Holland, New York, 1978.
15. P. Grisvard. *Elliptic Problems in Nonsmooth Domains.* Pitman, Boston, 1985.
16. G. Haase, U. Langer, A. Meyer, and S. V. Nepomnyaschik. Hierarchical extension operators and local multigrid methods in domain decomposition preconditioners. *East–West J. Numer. Math.*, 2:173–193, 1994.
17. L. Hörmander. *Linear Partial Differential Operators.* Springer-Verlag, New York, 1963.
18. U. Kotyczka and Peter Oswald. Piecewise linear prewavelets of small support. In C. Chui and L.L. Schumaker, editors, *Approximation Theory VIII, vol. 2 (wavelets and multilevel approximation)*, pages 235–242, Singapore, 1995. World Scientific.
19. J.L. Lions and E. Magenes. *Non-Homogeneous Boundary Value Problems and Applications.* Springer-Verlag, New York, 1972.
20. R. Lorentz and Peter Oswal. Multilevel additive methods for elliptic finite element problems. In *Proceedings of the Domain Decomposition Conference held in Bergen, Norway, June 3–8*, 1996.
21. Frank Natterer. Error bounds for tikhonov regularization in hilbert scales. *Appl. Anal.*, 18:29–37, 1984.

22. S. V. Nepomnyaschikh. Optimal multilevel extension operators. Technical report, Technische Universität Chemnitz–Zwickau, Germany, 1995. Report SPC 95–3.
23. P. Oswald. On discrete norm estimates related to multilevel preconditioners in the finite element method. In P. Petrushev K. G. Ivanov and B. Sendov, editors, *Constructive Theory of Functions*, pages 203–214, Bulg. Acad. Sci., Sofia, 1992. Proc. Int. Conf. Varna.
24. Peter Oswald. *Multilevel Finite Element Approximation, Theory and Applications.* Teubner Stuttgart, 1994.
25. Rob Stevenson. Piecewise linear (pre-) wavelets on non–uniform meshes. Technical report, Department of Mathematics, University of Nijmegen, The Netherlands, 1995.
26. Rob Stevenson. A robust hierarchical basis preconditioner on general meshes. Technical report, Department of Mathematics, University of Nijmegen, The Netherlands, 1995.
27. U. Tautenhahn. Error estimates for regularization methods in hilbert scales. *SIAM J. Numer. Anal.*, 33:2120–2130, 1996.
28. Panayot S. Vassilevski and Junping Wang. Stabilizing the hierarchical basis by approximate wavelets, i: Theory. *Numer. Linear Alg. Appl.*, 4:103–126, 1997.
29. Panayot S. Vassilevski and Junping Wang. Stabilizing the hierarchical basis by approximate wavelets, ii: Implementation and numerical experiments. *SIAM J. Sci. Comput.*, 2000.
30. Xuejun Zhang. Multi-level additive Schwarz methods. Technical report, Courant Inst. Math. Sci., Dept. Comp. Sci. Rep, 1991.

List of Participants

1. BAEZA Antonio, Spain,
   anbaman@alumni.uv.es
2. BELDA Ana Maria, Spain,
   abelgar@alumni.uv.es
3. BERRONE Stefano, Italy,
   sberrone@calvino.polito.it
4. BERTOLUZZA Silvia, Italy,
   aivlis@dragon.ian.pv.cnr.it
5. BRAMBLE James, USA, (lecturer)
   bramble@math.tamu.edu
6. BRAUER Philipp, Germany,
   brauer@iam.uni-bonn.de
7. BUFFA Annalisa, Italy
   annalisa@dragon.ian.pv.cnr.it
8. BURSTEDDE Carsten, Germany
   bursted@iam.uni-bonn.de
9. CARFORA Maria Francesca, Italy,
   f.carfora@iam.na.cnr.it
10. CATINAS Daniela, Romania,
    Daniela.Catinas@math.utcluj.ro
11. COHEN Albert, France, (lecturer)
    cohen@ann.jussieu.fr
12. CANUTO Claudio, Italy, (editor)
    ccanuto@polito.it
13. CORDERO Elena, Italy,
    cordero@dm.unito.it
14. DAHMEN Wolfgang, Germany, (lecturer)
    dahmen@igpm.rwth-aachen.de
15. FALLETTA Silvia, Italy,
    falletta@dragon.ian.pv.cnr.it

16. FRYZLEWICZ Piotr, Poland,
p.z.fryzlewicz@bris.ac.uk
17. GHEORGHE Simona, Romania,
simona@mecsol.ro
18. GRAVEMEIER Volker, Germany,
gravem@statik.uni-stuttgard.de
19. HAUSENBLAS Erika, Austria,
erika.hausenblas@sbg.ac.at
20. HUND Andrea, Germany,
hund@statik.uni-stuttgard.de
21. JAECKEL Uwe, Germany,
jaekel@ccrl-nece.de
22. JUERGENS Markus, Germany,
juergens@igpm.rwth-aachen.de
23. JYLHAKALLIO Mervi, Finland,
Mervi.Jylhakallio@fi.abb.com
24. KONDRATYUK Yaroslav, Ukraine,
kondratyuk@ukrpost.net
25. LEE Kelvin, Singapore,
ecmlee@ntu.edu.sg
26. LELIEVRE Tony, France,
tony.lelievre@polytechnique.org
27. LENZ Stefano, Germany,
stefan.lenz@bv.tum.de
28. LOIX Fabrice, Belgium,
loix@mema.ucl.ac.be
29. MALENGIER Benny, Belgium,
bm@cage.rug.ac.be
30. MICHELETTI Stefano, Italy,
stefano.micheletti@polimi.it
31. MOMMER Mario, Germany,
mommer@igpm.rwth-aachen.de
32. NGUYEN Hoang, Netherlands,
nguyen@math.uu.nl
33. PEROTTO Simona, Italy,
simona@mate.polimi.it
34. ROHRIG Susanne, Germany,
roehrig@informatik.uni-bonn.de
35. RUSSO Alessandro, Italy,
russo@dragon.ian.pv.cnr.it
36. SALERNO Ginevra, Italy,
ginevra@uniroma3.it
37. SANGALLI Giancarlo, Italy,
sangalli@dimat.unipv.it

38. SANSALONE Vittorio, Italy,
vsansa@dsic.uniroma3.it
39. SERNA Susana, Spain,
Serna.Susana@uv.es
40. SHYSHKANOVA Ganna, Ukraine,
shganna@yahoo.com
41. TABACCO Anita, Italy,
tabacco@polito.it
42. VAN GOTHEM Nicolas, Belgium,
vangoeth@mema.ucl.ac.be
43. VERANI Marco, Italy,
verani@dragon.ian.pv.cnr.it

LIST OF C.I.M.E. SEMINARS

1954	1. Analisi funzionale	C.I.M.E
	2. Quadratura delle superficie e questioni connesse	"
	3. Equazioni differenziali non lineari	"
1955	4. Teorema di Riemann-Roch e questioni connesse	"
	5. Teoria dei numeri	"
	6. Topologia	"
	7. Teorie non linearizzate in elasticità, idrodinamica, aerodinamic	"
	8. Geometria proiettivo-differenziale	"
1956	9. Equazioni alle derivate parziali a caratteristiche reali	"
	10. Propagazione delle onde elettromagnetiche	"
	11. Teoria della funzioni di più variabili complesse e delle funzioni automorfe	"
1957	12. Geometria aritmetica e algebrica (2 vol.)	"
	13. Integrali singolari e questioni connesse	"
	14. Teoria della turbolenza (2 vol.)	"
1958	15. Vedute e problemi attuali in relatività generale	"
	16. Problemi di geometria differenziale in grande	"
	17. Il principio di minimo e le sue applicazioni alle equazioni funzionali	"
1959	18. Induzione e statistica	"
	19. Teoria algebrica dei meccanismi automatici (2 vol.)	"
	20. Gruppi, anelli di Lie e teoria della coomologia	"
1960	21. Sistemi dinamici e teoremi ergodici	"
	22. Forme differenziali e loro integrali	"
1961	23. Geometria del calcolo delle variazioni (2 vol.)	"
	24. Teoria delle distribuzioni	"
	25. Onde superficiali	"
1962	26. Topologia differenziale	"
	27. Autovalori e autosoluzioni	"
	28. Magnetofluidodinamica	"
1963	29. Equazioni differenziali astratte	"
	30. Funzioni e varietà complesse	"
	31. Proprietà di media e teoremi di confronto in Fisica Matematica	"

1964	32. Relatività generale	C.I.M.E
	33. Dinamica dei gas rarefatti	"
	34. Alcune questioni di analisi numerica	"
	35. Equazioni differenziali non lineari	"
1965	36. Non-linear continuum theories	"
	37. Some aspects of ring theory	"
	38. Mathematical optimization in economics	"
1966	39. Calculus of variations	Ed. Cremonese, Firenze
	40. Economia matematica	"
	41. Classi caratteristiche e questioni connesse	"
	42. Some aspects of diffusion theory	"
1967	43. Modern questions of celestial mechanics	"
	44. Numerical analysis of partial differential equations	"
	45. Geometry of homogeneous bounded domains	"
1968	46. Controllability and observability	"
	47. Pseudo-differential operators	"
	48. Aspects of mathematical logic	"
1969	49. Potential theory	"
	50. Non-linear continuum theories in mechanics and physics and their applications	"
	51. Questions of algebraic varieties	"
1970	52. Relativistic fluid dynamics	"
	53. Theory of group representations and Fourier analysis	"
	54. Functional equations and inequalities	"
	55. Problems in non-linear analysis	"
1971	56. Stereodynamics	"
	57. Constructive aspects of functional analysis (2 vol.)	"
	58. Categories and commutative algebra	"
1972	59. Non-linear mechanics	"
	60. Finite geometric structures and their applications	"
	61. Geometric measure theory and minimal surfaces	"
1973	62. Complex analysis	"
	63. New variational techniques in mathematical physics	"
	64. Spectral analysis	"
1974	65. Stability problems	"
	66. Singularities of analytic spaces	"
	67. Eigenvalues of non linear problems	"
1975	68. Theoretical computer sciences	"
	69. Model theory and applications	"
	70. Differential operators and manifolds	"
1976	71. Statistical Mechanics	Ed. Liguori, Napoli
	72. Hyperbolicity	"
	73. Differential topology	"
1977	74. Materials with memory	"
	75. Pseudodifferential operators with applications	"
	76. Algebraic surfaces	"

1978	77. Stochastic differential equations		Ed. Liguori, Napoli & Birkhäuser
	78. Dynamical systems		"
1979	79. Recursion theory and computational complexity		"
	80. Mathematics of biology		"
1980	81. Wave propagation		"
	82. Harmonic analysis and group representations		"
	83. Matroid theory and its applications		"
1981	84. Kinetic Theories and the Boltzmann Equation	(LNM 1048)	Springer-Verlag
	85. Algebraic Threefolds	(LNM 947)	"
	86. Nonlinear Filtering and Stochastic Control	(LNM 972)	"
1982	87. Invariant Theory	(LNM 996)	"
	88. Thermodynamics and Constitutive Equations	(LN Physics 228)	"
	89. Fluid Dynamics	(LNM 1047)	"
1983	90. Complete Intersections	(LNM 1092)	"
	91. Bifurcation Theory and Applications	(LNM 1057)	"
	92. Numerical Methods in Fluid Dynamics	(LNM 1127)	"
1984	93. Harmonic Mappings and Minimal Immersions	(LNM 1161)	"
	94. Schrödinger Operators	(LNM 1159)	"
	95. Buildings and the Geometry of Diagrams	(LNM 1181)	"
1985	96. Probability and Analysis	(LNM 1206)	"
	97. Some Problems in Nonlinear Diffusion	(LNM 1224)	"
	98. Theory of Moduli	(LNM 1337)	"
1986	99. Inverse Problems	(LNM 1225)	"
	100. Mathematical Economics	(LNM 1330)	"
	101. Combinatorial Optimization	(LNM 1403)	"
1987	102. Relativistic Fluid Dynamics	(LNM 1385)	"
	103. Topics in Calculus of Variations	(LNM 1365)	"
1988	104. Logic and Computer Science	(LNM 1429)	"
	105. Global Geometry and Mathematical Physics	(LNM 1451)	"
1989	106. Methods of nonconvex analysis	(LNM 1446)	"
	107. Microlocal Analysis and Applications	(LNM 1495)	"
1990	108. Geometric Topology: Recent Developments	(LNM 1504)	"
	109. H_∞ Control Theory	(LNM 1496)	"
	110. Mathematical Modelling of Industrial Processes	(LNM 1521)	"
1991	111. Topological Methods for Ordinary Differential Equations	(LNM 1537)	"
	112. Arithmetic Algebraic Geometry	(LNM 1553)	"
	113. Transition to Chaos in Classical and Quantum Mechanics	(LNM 1589)	"
1992	114. Dirichlet Forms	(LNM 1563)	"
	115. D-Modules, Representation Theory, and Quantum Groups	(LNM 1565)	"
	116. Nonequilibrium Problems in Many-Particle Systems	(LNM 1551)	"

1993	117. Integrable Systems and Quantum Groups	(LNM 1620)	Springer-Verlag
	118. Algebraic Cycles and Hodge Theory	(LNM 1594)	"
	119. Phase Transitions and Hysteresis	(LNM 1584)	"
1994	120. Recent Mathematical Methods in Nonlinear Wave Propagation	(LNM 1640)	"
	121. Dynamical Systems	(LNM 1609)	"
	122. Transcendental Methods in Algebraic Geometry	(LNM 1646)	"
1995	123. Probabilistic Models for Nonlinear PDE's	(LNM 1627)	"
	124. Viscosity Solutions and Applications	(LNM 1660)	"
	125. Vector Bundles on Curves. New Directions	(LNM 1649)	"
1996	126. Integral Geometry, Radon Transforms and Complex Analysis	(LNM 1684)	"
	127. Calculus of Variations and Geometric Evolution Problems	(LNM 1713)	"
	128. Financial Mathematics	(LNM 1656)	"
1997	129. Mathematics Inspired by Biology	(LNM 1714)	"
	130. Advanced Numerical Approximation of Nonlinear Hyperbolic Equations	(LNM 1697)	"
	131. Arithmetic Theory of Elliptic Curves	(LNM 1716)	"
	132. Quantum Cohomology	(LNM 1776)	"
1998	133. Optimal Shape Design	(LNM 1740)	"
	134. Dynamical Systems and Small Divisors	(LNM 1784)	"
	135. Mathematical Problems in Semiconductor Physics	(LNM 1823)	"
	136. Stochastic PDE's and Kolmogorov Equations in Infinite Dimension	(LNM 1715)	"
	137. Filtration in Porous Media and Industrial Applications	(LNM 1734)	"
1999	138. Computational Mathematics driven by Industrial Applications	(LNM 1739)	"
	139. Iwahori-Hecke Algebras and Representation Theory	(LNM 1804)	"
	140. Theory and Applications of Hamiltonian Dynamics	to appear	"
	141. Global Theory of Minimal Surfaces in Flat Spaces	(LNM 1775)	"
	142. Direct and Inverse Methods in Solving Nonlinear Evolution Equations	to appear	"
2000	143. Dynamical Systems	(LNM 1822)	"
	144. Diophantine Approximation	(LNM 1819)	"
	145. Mathematical Aspects of Evolving Interfaces	(LNM 1812)	"
	146. Mathematical Methods for Protein Structure	to appear	"
	147. Noncommutative Geometry	to appear	"
2001	148. Topological Fluid Mechanics	to appear	"
	149. Spatial Stochastic Processes	(LNM 1802)	"
	150. Optimal Transportation and Applications	(LNM 1813)	"
	151. Multiscale Problems and Methods in Numerical Simulations	(LNM 1825)	"
2002	152. Real methods in Complex and CR Geometry	to appear	"
	153. Analytic Number Theory	to appear	"
	154. Imaging	to appear	"

2003	155. Stochastik Methods in Finance	announced
	156. Hyperbolik Systems of Balance Laws	announced
	157. Sympletic 4-Manifolds and Algebraic Surfaces	announced
	158. Mathematical Foundation of Turbulent Viscous Flows	announced
2004	159. Representation Theory and Complex Analysis	announced
	160. Nonlinear and Optimal Control Theory	announced
	161. Stochastic Geometry	announced

Fondazione C.I.M.E.

Centro Internazionale Matematico Estivo
International Mathematical Summer Center
http://www.math.unifi.it/~cime
cime@math.unifi.it

2004 COURSES LIST

Representation Theory and Complex Analysis

June 10–17, Venezia

Course Directors:

Prof. Enrico Casadio Tarabusi (Università di Roma "La Sapienza")
Prof. Andrea D'Agnolo (Università di Padova)
Prof. Massimo A. Picardello (Università di Roma "Tor Vergata")

Nonlinear and Optimal Control Theory

June 21–29, Cetraro (Cosenza)

Course Directors:

Prof. Paolo Nistri (Università di Siena)
Prof. Gianna Stefani (Università di Firenze)

Stochastic Geometry

September 13–18, Martina Franca (Taranto)

Course Director:

Prof. W. Weil (Univ. of Karlsruhe, Karlsruhe, Germany)

Lecture Notes in Mathematics

For information about Vols. 1–1648
please contact your bookseller or Springer-Verlag

Vol. 1649: S. Kumar, G. Laumon, U. Stuhler, Vector Bundles on Curves – New Directions. Cetraro, 1995. Editor: M. S. Narasimhan. VII, 193 pages. 1997.

Vol. 1650: J. Wildeshaus, Realizations of Polylogarithms. XI, 343 pages. 1997.

Vol. 1651: M. Drmota, R. F. Tichy, Sequences, Discrepancies and Applications. XIII, 503 pages. 1997.

Vol. 1652: S. Todorcevic, Topics in Topology. VIII, 153 pages. 1997.

Vol. 1653: R. Benedetti, C. Petronio, Branched Standard Spines of 3-manifolds. VIII, 132 pages. 1997.

Vol. 1654: R. W. Ghrist, P. J. Holmes, M. C. Sullivan, Knots and Links in Three-Dimensional Flows. X, 208 pages. 1997.

Vol. 1655: J. Azéma, M. Emery, M. Yor (Eds.), Séminaire de Probabilités XXXI. VIII, 329 pages. 1997.

Vol. 1656: B. Biais, T. Björk, J. Cvitanic, N. El Karoui, E. Jouini, J. C. Rochet, Financial Mathematics. Bressanone, 1996. Editor: W. J. Runggaldier. VII, 316 pages. 1997.

Vol. 1657: H. Reimann, The semi-simple zeta function of quaternionic Shimura varieties. IX, 143 pages. 1997.

Vol. 1658: A. Pumarino, J. A. Rodrýguez, Coexistence and Persistence of Strange Attractors. VIII, 195 pages. 1997.

Vol. 1659: V, Kozlov, V. Maz'ya, Theory of a Higher-Order Sturm-Liouville Equation. XI, 140 pages. 1997.

Vol. 1660: M. Bardi, M. G. Crandall, L. C. Evans, H. M. Soner, P. E. Souganidis, Viscosity Solutions and Applications. Montecatini Terme, 1995. Editors: I. Capuzzo Dolcetta, P. L. Lions. IX, 259 pages. 1997.

Vol. 1661: A. Tralle, J. Oprea, Symplectic Manifolds with no Kähler Structure. VIII, 207 pages. 1997.

Vol. 1662: J. W. Rutter, Spaces of Homotopy Self-Equivalences – A Survey. IX, 170 pages. 1997.

Vol. 1663: Y. E. Karpeshina; Perturbation Theory for the Schrödinger Operator with a Periodic Potential. VII, 352 pages. 1997.

Vol. 1664: M. Väth, Ideal Spaces. V, 146 pages. 1997.

Vol. 1665: E. Giné, G. R. Grimmett, L. Saloff-Coste, Lectures on Probability Theory and Statistics 1996. Editor: P. Bernard. X, 424 pages, 1997.

Vol. 1666: M. van der Put, M. F. Singer, Galois Theory of Difference Equations. VII, 179 pages. 1997.

Vol. 1667: J. M. F. Castillo, M. González, Three-space Problems in Banach Space Theory. XII, 267 pages. 1997.

Vol. 1668: D. B. Dix, Large-Time Behavior of Solutions of Linear Dispersive Equations. XIV, 203 pages. 1997.

Vol. 1669: U. Kaiser, Link Theory in Manifolds. XIV, 167 pages. 1997.

Vol. 1670: J. W. Neuberger, Sobolev Gradients and Differential Equations. VIII, 150 pages. 1997.

Vol. 1671: S. Bouc, Green Functors and *G*-sets. VII, 342 pages. 1997.

Vol. 1672: S. Mandal, Projective Modules and Complete Intersections. VIII, 114 pages. 1997.

Vol. 1673: F. D. Grosshans, Algebraic Homogeneous Spaces and Invariant Theory. VI, 148 pages. 1997.

Vol. 1674: G. Klaas, C. R. Leedham-Green, W. Plesken, Linear Pro-*p*-Groups of Finite Width. VIII, 115 pages. 1997.

Vol. 1675: J. E. Yukich, Probability Theory of Classical Euclidean Optimization Problems. X, 152 pages. 1998.

Vol. 1676: P. Cembranos, J. Mendoza, Banach Spaces of Vector-Valued Functions. VIII, 118 pages. 1997.

Vol. 1677: N. Proskurin, Cubic Metaplectic Forms and Theta Functions. VIII, 196 pages. 1998.

Vol. 1678: O. Krupková, The Geometry of Ordinary Variational Equations. X, 251 pages. 1997.

Vol. 1679: K.-G. Grosse-Erdmann, The Blocking Technique. Weighted Mean Operators and Hardy's Inequality. IX, 114 pages. 1998.

Vol. 1680: K.-Z. Li, F. Oort, Moduli of Supersingular Abelian Varieties. V, 116 pages. 1998.

Vol. 1681: G. J. Wirsching, The Dynamical System Generated by the 3n+1 Function. VII, 158 pages. 1998.

Vol. 1682: H.-D. Alber, Materials with Memory. X, 166 pages. 1998.

Vol. 1683: A. Pomp, The Boundary-Domain Integral Method for Elliptic Systems. XVI, 163 pages. 1998.

Vol. 1684: C. A. Berenstein, P. F. Ebenfelt, S. G. Gindikin, S. Helgason, A. E. Tumanov, Integral Geometry, Radon Transforms and Complex Analysis. Firenze, 1996. Editors: E. Casadio Tarabusi, M. A. Picardello, G. Zampieri. VII, 160 pages. 1998.

Vol. 1685: S. König, A. Zimmermann, Derived Equivalences for Group Rings. X, 146 pages. 1998.

Vol. 1686: J. Azéma, M. Émery, M. Ledoux, M. Yor (Eds.), Séminaire de Probabilités XXXII. VI, 440 pages. 1998.

Vol. 1687: F. Bornemann, Homogenization in Time of Singularly Perturbed Mechanical Systems. XII, 156 pages. 1998.

Vol. 1688: S. Assing, W. Schmidt, Continuous Strong Markov Processes in Dimension One. XII, 137 page. 1998.

Vol. 1689: W. Fulton, P. Pragacz, Schubert Varieties and Degeneracy Loci. XI, 148 pages. 1998.

Vol. 1690: M. T. Barlow, D. Nualart, Lectures on Probability Theory and Statistics. Editor: P. Bernard. VIII, 237 pages. 1998.

Vol. 1691: R. Bezrukavnikov, M. Finkelberg, V. Schechtman, Factorizable Sheaves and Quantum Groups. X, 282 pages. 1998.

Vol. 1692: T. M. W. Eyre, Quantum Stochastic Calculus and Representations of Lie Superalgebras. IX, 138 pages. 1998.

Vol. 1694: A. Braides, Approximation of Free-Discontinuity Problems. XI, 149 pages. 1998.

Vol. 1695: D. J. Hartfiel, Markov Set-Chains. VIII, 131 pages. 1998.

Vol. 1696: E. Bouscaren (Ed.): Model Theory and Algebraic Geometry. XV, 211 pages. 1998.

Vol. 1697: B. Cockburn, C. Johnson, C.-W. Shu, E. Tadmor, Advanced Numerical Approximation of Nonlinear Hyperbolic Equations. Cetraro, Italy, 1997. Editor: A. Quarteroni. VII, 390 pages. 1998.

Vol. 1698: M. Bhattacharjee, D. Macpherson, R. G. Möller, P. Neumann, Notes on Infinite Permutation Groups. XI, 202 pages. 1998.

Vol. 1699: A. Inoue,Tomita-Takesaki Theory in Algebras of Unbounded Operators. VIII, 241 pages. 1998.

Vol. 1700: W. A. Woyczyński, Burgers-KPZ Turbulence, XI, 318 pages. 1998.

Vol. 1701: Ti-Jun Xiao, J. Liang, The Cauchy Problem of Higher Order Abstract Differential Equations, XII, 302 pages. 1998.

Vol. 1702: J. Ma, J. Yong, Forward-Backward Stochastic Differential Equations and Their Applications. XIII, 270 pages. 1999.

Vol. 1703: R. M. Dudley, R. Norvaiša, Differentiability of Six Operators on Nonsmooth Functions and p-Variation. VIII, 272 pages. 1999.

Vol. 1704: H. Tamanoi, Elliptic Genera and Vertex Operator Super-Algebras. VI, 390 pages. 1999.

Vol. 1705: I. Nikolaev, E. Zhuzhoma, Flows in 2-dimensional Manifolds. XIX, 294 pages. 1999.

Vol. 1706: S. Yu. Pilyugin, Shadowing in Dynamical Systems. XVII, 271 pages. 1999.

Vol. 1707: R. Pytlak, Numerical Methods for Optimal Control Problems with State Constraints. XV, 215 pages. 1999.

Vol. 1708: K. Zuo, Representations of Fundamental Groups of Algebraic Varieties. VII, 139 pages. 1999.

Vol. 1709: J. Azéma, M. Émery, M. Ledoux, M. Yor (Eds), Séminaire de Probabilités XXXIII. VIII, 418 pages. 1999.

Vol. 1710: M. Koecher, The Minnesota Notes on Jordan Algebras and Their Applications. IX, 173 pages. 1999.

Vol. 1711: W. Ricker, Operator Algebras Generated by Commuting Projećtions: A Vector Measure Approach. XVII, 159 pages. 1999.

Vol. 1712: N. Schwartz, J. J. Madden, Semi-algebraic Function Rings and Reflectors of Partially Ordered Rings. XI, 279 pages. 1999.

Vol. 1713: F. Bethuel, G. Huisken, S. Müller, K. Steffen, Calculus of Variations and Geometric Evolution Problems. Cetraro, 1996. Editors: S. Hildebrandt, M. Struwe. VII, 293 pages. 1999.

Vol. 1714: O. Diekmann, R. Durrett, K. P. Hadeler, P. K. Maini, H. L. Smith, Mathematics Inspired by Biology. Martina Franca, 1997. Editors: V. Capasso, O. Diekmann. VII, 268 pages. 1999.

Vol. 1715: N. V. Krylov, M. Röckner, J. Zabczyk, Stochastic PDE's and Kolmogorov Equations in Infinite Dimensions. Cetraro, 1998. Editor: G. Da Prato. VIII, 239 pages. 1999.

Vol. 1716: J. Coates, R. Greenberg, K. A. Ribet, K. Rubin, Arithmetic Theory of Elliptic Curves. Cetraro, 1997. Editor: C. Viola. VIII, 260 pages. 1999.

Vol. 1717: J. Bertoin, F. Martinelli, Y. Peres, Lectures on Probability Theory and Statistics. Saint-Flour, 1997. Editor: P. Bernard. IX, 291 pages. 1999.

Vol. 1718: A. Eberle, Uniqueness and Non-Uniqueness of Semigroups Generated by Singular Diffusion Operators. VIII, 262 pages. 1999.

Vol. 1719: K. R. Meyer, Periodic Solutions of the N-Body Problem. IX, 144 pages. 1999.

Vol. 1720: D. Elworthy, Y. Le Jan, X-M. Li, On the Geometry of Diffusion Operators and Stochastic Flows. IV, 118 pages. 1999.

Vol. 1721: A. Iarrobino, V. Kanev, Power Sums, Gorenstein Algebras, and Determinantal Loci. XXVII, 345 pages. 1999.

Vol. 1722: R. McCutcheon, Elemental Methods in Ergodic Ramsey Theory. VI, 160 pages. 1999.

Vol. 1723: J. P. Croisille, C. Lebeau, Diffraction by an Immersed Elastic Wedge. VI, 134 pages. 1999.

Vol. 1724: V. N. Kolokoltsov, Semiclassical Analysis for Diffusions and Stochastic Processes. VIII, 347 pages. 2000.

Vol. 1725: D. A. Wolf-Gladrow, Lattice-Gas Cellular Automata and Lattice Boltzmann Models. IX, 308 pages. 2000.

Vol. 1726: V. Marić, Regular Variation and Differential Equations. X, 127 pages. 2000.

Vol. 1727: P. Kravanja M. Van Barel, Computing the Zeros of Analytic Functions. VII, 111 pages. 2000.

Vol. 1728: K. Gatermann Computer Algebra Methods for Equivariant Dynamical Systems. XV, 153 pages. 2000.

Vol. 1729: J. Azéma, M. Émery, M. Ledoux, M. Yor Séminaire de Probabilités XXXIV. VI, 431 pages. 2000.

Vol. 1730: S. Graf, H. Luschgy, Foundations of Quantization for Probability Distributions. X, 230 pages. 2000.

Vol. 1731: T. Hsu, Quilts: Central Extensions, Braid Actions, and Finite Groups. XII, 185 pages. 2000.

Vol. 1732: K. Keller, Invariant Factors, Julia Equivalences and the (Abstract) Mandelbrot Set. X, 206 pages. 2000.

Vol. 1733: K. Ritter, Average-Case Analysis of Numerical Problems. IX, 254 pages. 2000.

Vol. 1734: M. Espedal, A. Fasano, A. Mikelić, Filtration in Porous Media and Industrial Applications. Cetraro 1998. Editor: A. Fasano. 2000.

Vol. 1735: D. Yafaev, Scattering Theory: Some Old and New Problems. XVI, 169 pages. 2000.

Vol. 1736: B. O. Turesson, Nonlinear Potential Theory and Weighted Sobolev Spaces. XIV, 173 pages. 2000.

Vol. 1737: S. Wakabayashi, Classical Microlocal Analysis in the Space of Hyperfunctions. VIII, 367 pages. 2000.

Vol. 1738: M. Émery, A. Nemirovski, D. Voiculescu, Lectures on Probability Theory and Statistics. XI, 356 pages. 2000.

Vol. 1739: R. Burkard, P. Deuflhard, A. Jameson, J.-L. Lions, G. Strang, Computational Mathematics Driven by Industrial Problems. Martina Franca, 1999. Editors: V. Capasso, H. Engl, J. Periaux. VII, 418 pages. 2000.

Vol. 1740: B. Kawohl, O. Pironneau, L. Tartar, J.-P. Zolesio, Optimal Shape Design. Tróia, Portugal 1999. Editors: A. Cellina, A. Ornelas. IX, 388 pages. 2000.

Vol. 1741: E. Lombardi, Oscillatory Integrals and Phenomena Beyond all Algebraic Orders. XV, 413 pages. 2000.

Vol. 1742: A. Unterberger, Quantization and Nonholomorphic Modular Forms.VIII, 253 pages. 2000.

Vol. 1743: L. Habermann, Riemannian Metrics of Constant Mass and Moduli Spaces of Conformal Structures. XII, 116 pages. 2000.

Vol. 1744: M. Kunze, Non-Smooth Dynamical Systems. X, 228 pages. 2000.

Vol. 1745: V. D. Milman, G. Schechtman, Geometric Aspects of Functional Analysis. Israel Seminar 1999-2000. VIII, 289 pages. 2000.

Vol. 1746: A. Degtyarev, I. Itenberg, V. Kharlamov, Real Enriques Surfaces. XVI, 259 pages. 2000.

Vol. 1747: L. W. Christensen, Gorenstein Dimensions. VIII, 204 pages. 2000.

Vol. 1748: M. Ruzicka, Electrorheological Fluids: Modeling and Mathematical Theory. XV, 176 pages. 2001.

Vol. 1749: M. Fuchs, G. Seregin, Variational Methods for Problems from Plasticity Theory and for Generalized Newtonian Fluids. VI, 269 pages. 2001.

Vol. 1750: B. Conrad, Grothendieck Duality and Base Change. X, 296 pages. 2001.

Vol. 1751: N. J. Cutland, Loeb Measures in Practice: Recent Advances. XI, 111 pages. 2001.

Vol. 1752: Y. V. Nesterenko, P. Philippon, Introduction to Algebraic Independence Theory. XIII, 256 pages. 2001.

Vol. 1753: A. I. Bobenko, U. Eitner, Painlevé Equations in the Differential Geometry of Surfaces. VI, 120 pages. 2001.

Vol. 1754: W. Bertram, The Geometry of Jordan and Lie Structures. XVI, 269 pages. 2001.

Vol. 1755: J. Azéma, M. Émery, M. Ledoux, M. Yor, Séminaire de Probabilités XXXV. VI, 427 pages. 2001.

Vol. 1756: P. E. Zhidkov, Korteweg de Vries and Nonlinear Schrödinger Equations: Qualitative Theory. VII, 147 pages. 2001.

Vol. 1757: R. R. Phelps, Lectures on Choquet's Theorem. VII, 124 pages. 2001.

Vol. 1758: N. Monod, Continuous Bounded Cohomology of Locally Compact Groups. X, 214 pages. 2001.

Vol. 1759: Y. Abe, K. Kopfermann, Toroidal Groups. VIII, 133 pages. 2001.

Vol. 1760: D. Filipović, Consistency Problems for Heath-Jarrow-Morton Interest Rate Models. VIII, 134 pages. 2001.

Vol. 1761: C. Adelmann, The Decomposition of Primes in Torsion Point Fields. VI, 142 pages. 2001.

Vol. 1762: S. Cerrai, Second Order PDE's in Finite and Infinite Dimension. IX, 330 pages. 2001.

Vol. 1763: J.-L. Loday, A. Frabetti, F. Chapoton, F. Goichot, Dialgebras and Related Operads. IV, 132 pages. 2001.

Vol. 1764: A. Cannas da Silva, Lectures on Symplectic Geometry. XII, 217 pages. 2001.

Vol. 1765: T. Kerler, V. V. Lyubashenko, Non-Semisimple Topological Quantum Field Theories for 3-Manifolds with Corners. VI, 379 pages. 2001.

Vol. 1766: H. Hennion, L. Hervé, Limit Theorems for Markov Chains and Stochastic Properties of Dynamical Systems by Quasi-Compactness. VIII, 145 pages. 2001.

Vol. 1767: J. Xiao, Holomorphic Q Classes. VIII, 112 pages. 2001.

Vol. 1768: M.J. Pflaum, Analytic and Geometric Study of Stratified Spaces. VIII, 230 pages. 2001.

Vol. 1769: M. Alberich-Carramiñana, Geometry of the Plane Cremona Maps. XVI, 257 pages. 2002.

Vol. 1770: H. Gluesing-Luerssen, Linear Delay-Differential Systems with Commensurate Delays: An Algebraic Approach. VIII, 176 pages. 2002.

Vol. 1771: M. Émery, M. Yor, Séminaire de Probabilités 1967-1980. A Selection in Martingale Theory. IX, 553 pages. 2002.

Vol. 1772: F. Burstall, D. Ferus, K. Leschke, F. Pedit, U. Pinkall, Conformal Geometry of Surfaces in S^4. VII, 89 pages. 2002.

Vol. 1773: Z. Arad, M. Muzychuk, Standard Integral Table Algebras Generated by a Non-real Element of Small Degree. X, 126 pages. 2002.

Vol. 1774: V. Runde, Lectures on Amenability. XIV, 296 pages. 2002.

Vol. 1775: W. H. Meeks, A. Ros, H. Rosenberg, The Global Theory of Minimal Surfaces in Flat Spaces. Martina Franca 1999. Editor: G. P. Pirola. X, 117 pages. 2002.

Vol. 1776: K. Behrend, C. Gomez, V. Tarasov, G. Tian, Quantum Comohology. Cetraro 1997. Editors: P. de Bartolomeis, B. Dubrovin, C. Reina. VIII, 319 pages. 2002.

Vol. 1777: E. García-Río, D. N. Kupeli, R. Vázquez-Lorenzo, Osserman Manifolds in Semi-Riemannian Geometry. XII, 166 pages. 2002.

Vol. 1778: H. Kiechle, Theory of K-Loops. X, 186 pages. 2002.

Vol. 1779: I. Chueshov, Monotone Random Systems. VIII, 234 pages. 2002.

Vol. 1780: J. H. Bruinier, Borcherds Products on O(2,1) and Chern Classes of Heegner Divisors. VIII, 152 pages. 2002.

Vol. 1781: E. Bolthausen, E. Perkins, A. van der Vaart, Lectures on Probability Theory and Statistics. Ecole d' Eté de Probabilités de Saint-Flour XXIX-1999. Editor: P. Bernard. VIII, 466 pages. 2002.

Vol. 1782: C.-H. Chu, A. T.-M. Lau, Harmonic Functions on Groups and Fourier Algebras. VII, 100 pages. 2002.

Vol. 1783: L. Grüne, Asymptotic Behavior of Dynamical and Control Systems under Perturbation and Discretization. IX, 231 pages. 2002.

Vol. 1784: L.H. Eliasson, S. B. Kuksin, S. Marmi, J.-C. Yoccoz, Dynamical Systems and Small Divisors. Cetraro, Italy 1998. Editors: S. Marmi, J.-C. Yoccoz. VIII, 199 pages. 2002.

Vol. 1785: J. Arias de Reyna, Pointwise Convergence of Fourier Series. XVIII, 175 pages. 2002.

Vol. 1786: S. D. Cutkosky, Monomialization of Morphisms from 3-Folds to Surfaces. V, 235 pages. 2002.

Vol. 1787: S. Caenepeel, G. Militaru, S. Zhu, Frobenius and Separable Functors for Generalized Module Categories and Nonlinear Equations. XIV, 354 pages. 2002.

Vol. 1788: A. Vasil'ev, Moduli of Families of Curves for Conformal and Quasiconformal Mappings.IX, 211 pages. 2002.

Vol. 1789: Y. Sommerhäuser, Yetter-Drinfel'd Hopf algebras over groups of prime order. V, 157 pages. 2002.

Vol. 1790: X. Zhan, Matrix Inequalities. VII, 116 pages. 2002.

Vol. 1791: M. Knebusch, D. Zhang, Manis Valuations and Prüfer Extensions I: A new Chapter in Commutative Algebra. VI, 267 pages. 2002.

Vol. 1792: D. D. Ang, R. Gorenflo, V. K. Le, D. D. Trong, Moment Theory and Some Inverse Problems in Potential Theory and Heat Conduction. VIII, 183 pages. 2002.

Vol. 1793: J. Cortés Monforte, Geometric, Control and Numerical Aspects of Nonholonomic Systems. XV, 219 pages. 2002.

Vol. 1794: N. Pytheas Fogg, Substitution in Dynamics, Arithmetics and Combinatorics. Editors: V. Berthé, S. Ferenczi, C. Mauduit, A. Siegel. XVII, 402 pages. 2002.

Vol. 1795: H. Li, Filtered-Graded Transfer in Using Non-commutative Gröbner Bases. IX, 197 pages. 2002.

Vol. 1796: J.M. Melenk, hp-Finite Element Methods for Singular Perturbations. XIV, 318 pages. 2002.

Vol. 1797: B. Schmidt, Characters and Cyclotomic Fields in Finite Geometry. VIII, 100 pages. 2002.

Vol. 1798: W.M. Oliva, Geometric Mechanics. XI, 270 pages. 2002.

Vol. 1799: H. Pajot, Analytic Capacity, Rectifiability, Menger Curvature and the Cauchy Integral. XII,119 pages. 2002.

Vol. 1800: O. Gabber, L. Ramero, Almost Ring Theory. VI, 307 pages. 2003.

Vol. 1801: J. Azéma, M. Émery, M. Ledoux, M. Yor, Séminaire de Probabilités XXXVI. VIII, 499 pages. 2003.

Vol. 1802: V. Capasso, E. Merzbach, B.G. Ivanoff, M. Dozzi, R. Dalang, T. Mountford, Topics in Spatial Stochastic Processes. Martina Franca, Italy 2001. Editor: E. Merzbach. VIII, 253 pages. 2003.

Vol. 1803: G. Dolzmann, Variational Methods for Crystalline Microstructure - Analysis and Computation. VIII, 212 pages. 2003.

Vol. 1804: I. Cherednik, Ya. Markov, R. Howe, G. Lusztig, Iwahori-Hecke Algebras and their Representation Theory. Martina Franca, Italy 1999. Editors: V. Baldoni, D. Barbasch. X, 103 pages. 2003.

Vol. 1805: F. Cao, Geometric Curve Evolution and Image Processing. X, 187 pages. 2003.

Vol. 1806: H. Broer, I. Hoveijn. G. Lunther, G. Vegter, Bifurcations in Hamiltonian Systems. Computing Singularities by Gröbner Bases. XIV, 169 pages. 2003.

Vol. 1807: V. D. Milman, G. Schechtman, Geometric Aspects of Functional Analysis. Israel Seminar 2000-2002. VIII, 429 pages. 2003.

Vol. 1808: W. Schindler, Measures with Symmetry Properties.IX, 167 pages. 2003.

Vol. 1809: O. Steinbach, Stability Estimates for Hybrid Coupled Domain Decomposition Methods. VI, 120 pages. 2003.

Vol. 1810: J. Wengenroth, Derived Functors in Functional Analysis. VIII, 134 pages. 2003.

Vol. 1811: J. Stevens, Deformations of Singularities. VII, 157 pages. 2003.

Vol. 1812: L. Ambrosio, K. Deckelnick, G. Dziuk, M. Mimura, V. A. Solonnikov, H. M. Soner, Mathematical Aspects of Evolving Interfaces. Madeira, Funchal, Portugal 2000. Editors: P. Colli, J. F. Rodrigues. X, 237 pages. 2003.

Vol. 1813: L. Ambrosio, L. A. Caffarelli, Y. Brenier, G. Buttazzo, C. Villani, Optimal Transportation and its Applications. Martina Franca, Italy 2001. Editors: L. A. Caffarelli, S. Salsa. X, 164 pages. 2003.

Vol. 1814: P. Bank, F. Baudoin, H. Föllmer, L.C.G. Rogers, M. Soner, N. Touzi, Paris-Princeton Lectures on Mathematical Finance. X,172 pages. 2003.

Vol. 1815: A. M. Vershik (Ed.), Asymptotic Combinatorics with Applications to Mathematical Physics. St. Petersburg, Russia 2001. IX, 246 pages. 2003.

Vol. 1816: S. Albeverio, W. Schachermayer, M. Talagrand, Lectures on Probability Theory and Statistics. Ecole d'Eté de Probabilités de Saint-Flour XXX-2000. Editor: P. Bernard. VIII, 296 pages. 2003.

Vol. 1817: E. Koelink (Ed.), Orthogonal Polynomials and Special Functions. Leuven 2002. X, 249 pages. 2003.

Vol. 1818: M. Bildhauer, Convex Variational Problems with Linear, nearly Linear and/or Anisotropic Growth Conditions. X, 217 pages. 2003.

Vol. 1819: D. Masser, Yu. V. Nesterenko, H. P. Schlickewei, W. M. Schmidt, M. Waldschmidt, Diophantine Approximation. Cetraro, Italy 2000. Editors: F. Amoroso, U. Zannier. XI,353 pages. 2003.

Vol. 1820: F. Hiai, H. Kosaki, Means of Hilbert Space Operators. VIII, 148 pages. 2003.

Vol. 1821: S. Teufel, Adiabatic Perturbation Theory in Quantum Dynamics. VI, 242 pages. 2003.

Vol. 1822: S.-N. Chow, R. Conti, R. Johnson, J. Mallet-Paret, R. Nussbaum, Dynamical Systems. Cetraro, Italy 2000. Editors: J. W. Macki, P. Zecca. XII, 345 pages. 2003.

Vol. 1823: A. M. Anile, W. Allegretto, C. Ringhofer, Mathematical Problems in Semiconductor Physics. Cetraro, Italy 1998. Editor: A. M. Anile. X, 135 pages. 2003.

Vol. 1824: J. A. Navarro González, J. B. Sancho de Salas, $\mathcal{C}^\infty$ - Differentiable Spaces. XIII, 188 pages. 2003.

Vol. 1825: J. H. Bramble, A. Cohen, W. Dahmen, Multiscale Problems and Methods in Numerical Simulations, Martina Franca, Italy 2001. Editor: C. Canuto. XIII, 163 pages. 2003.

Vol. 1826: K. Dohmen, Improved Bonferroni Inequalities via Abstract Tubes. Inequalities and Identities of Inclusion-Exclusion Type. VIII, 113 pages, 2003.

Vol. 1827: K. M. Pilgrim, Combinations of Complex Dynamical Systems. X, 118 pages, 2003.

Vol. 1828: D. J. Green, Gröbner Bases and the Computation of Group Cohomology. XII, 138 pages, 2003.

Vol. 1829: E. Altman, B. Gaujal, A. Hordijk, Discrete-Event Control of Stochastic Networks: Multimodularity and Regularity. XIV, 313 pages, 2003.

Vol. 1830: M. I. Gil', Operator Functions and Localization of Spectra. XIV, 256 pages, 2003.

Vol. 1831: A. Connes, J. Cuntz, E. Guentner, N. Higson, J. E. Kaminker, Noncommutative Geometry, Martina Franca, Italy 2002. Editors: S. Doplicher, L. Longo. XV, 344 pages. 2003.

Recent Reprints and New Editions

Vol. 1200: V. D. Milman, G. Schechtman, Asymptotic Theory of Finite Dimensional Normed Spaces. 1986. – Corrected Second Printing. X, 156 pages. 2001.

Vol. 1471: M. Courtieu, A.A. Panchishkin, Non-Archimedean L-Functions and Arithmetical Siegel Modular Forms. – Second Edition. VII. 196 pages. 2003.

Vol. 1618: G. Pisier, Similarity Problems and Completely Bounded Maps. 1995 – Second, Expanded Edition VII, 198 pages. 2001.

Vol. 1629: J. D. Moore, Lectures on Seiberg-Witten Invariants. 1997 – Second Edition. VIII, 121 pages. 2001.

Vol. 1638: P. Vanhaecke, Integrable Systems in the realm of Algebraic Geometry. 1996 – Second Edition. X, 256 pages. 2001.

Vol. 1702: J. Ma, J. Yong, Forward-Backward Stochastic Differential Equations and Their Applications. 1999. – Corrected Second Printing. XIII, 270 pages. 2000.